Enhancing Deep Learning with Bayesian Inference

Create more powerful, robust deep learning systems with Bayesian deep learning in Python

Dr. Matt Benatan

Jochem Gietema

Dr. Marian Schneider

‹packt›

BIRMINGHAM—MUMBAI

Enhancing Deep Learning with Bayesian Inference

Group Product Manager: Ali Abidi
Senior Editor: David Sugarman
Technical Editor: Devanshi Ayare
Copy Editor: Safis Editing
Project Coordinator Farheen Fathima
Proofreader: Safis Editing
Indexer Rekha Nair
Production Designer: Jyoti Chauhan

First published: June 2023

Production reference: 1280623

Published by Packt Publishing Ltd.
Livery Place
35 Livery Street
Birmingham
B3 2PB, UK.

ISBN 978-1-80324-688-8

www.packtpub.com

Contributors

About the authors

Dr. Matt Benatan is a principal research scientist at Sonos, where he leads research into intelligent personalization systems. He has also been awarded a Simon Industrial Fellowship at the University of Manchester, where he collaborates on a variety of AI research projects. Matt obtained his PhD in audio-visual speech processing from the University of Leeds, after which he pursued a career in industry, conducting machine learning research across a range of domains including signal processing, materials discovery, and fraud detection. Matt previously co-authored Wiley's *Deep Learning for Physical Scientists*, and his key research interests currently include user-facing AI, optimization, and uncertainty estimation.

Matt is deeply grateful to his wife, Rebecca, for her love, patience, and support, and to his parents, Dan and Debby, for their tireless enthusiasm, guidance, and encouragement.

Jochem Gietema studied philosophy and law in Amsterdam but transitioned to machine learning after his studies. He currently works as an applied scientist at Onfido in London, where he has developed and deployed several patented solutions to production in the field of computer vision and anomaly detection. He is passionate about uncertainty estimation, interactive data visualizations, and solving real-world problems with machine learning.

Dr. Marian Schneider is an applied scientist in machine learning and computer vision. He received his PhD in computational visual neuroscience from the University of Maastricht. He has since transitioned from academia to industry, where he has developed and applied machine learning solutions to a wide range of products, from brain image segmentation to uncertainty estimation, to smarter image capturing on mobile phone devices.

Marian is grateful to his partner, Undine, who was very supportive of the book writing process and enabled many writing and working sessions on this book, especially during weekends.

About the reviewers

Neba Nfonsang is a data scientist and an instructor of data science and statistics at the University of Denver. He earned his PhD in research methods and statistics at the University of Denver. Neba has experience working both in academia and industry, and has taught 12 graduate courses, including data science and statistics courses. He has provided training and best practices instruction to companies around the world on advanced analytics and designing production-ready statistical and machine learning models.

Avijit K is an accomplished Chief Information Officer (CIO) with over 15 years of experience in the field of data science and artificial intelligence. He holds a PhD in computer science from a top-tier university and has completed several industry projects in machine learning, deep learning, computer vision, natural language processing, and related technologies. In his current role as CIO at a services company, Avi is responsible for overseeing the organization's technology operations, including data analytics, infrastructure, and software development.

Table of Contents

Preface

Over the last decade, the field of machine learning has taken great strides, and in so doing has captured the public's imagination. But it's crucial to remember that – as impressive as these algorithms are – they are not infallible. Through this book, we hope to provide an approachable introduction to how Bayesian inference can be leveraged within deep learning, giving the reader the tools to develop models that "know when they don't know." In so doing, you'll be able to develop more robust deep learning systems better suited to the demands of today's machine learning-based applications.

Who this book is for

This book is for researchers, developers, and engineers who work on the development and application of machine learning algorithms, and who want to start working with uncertainty-aware deep learning models.

What this book covers

Chapter 1, Bayesian Inference in the Age of Deep Learning, covers use cases and limitations of traditional deep learning methods.

Chapter 2, Fundamentals of Bayesian Inference, discusses Bayesian modeling and inference and explores gold-standard machine learning methods for Bayesian inference.

Chapter 3, Fundamentals of Deep Learning, introduces you to the main building blocks of deep learning models.

Chapter 4, Introducing Bayesian Deep Learning, combines the concepts introduced in *Chapter 2* and *Chapter 3* to discuss Bayesian deep learning.

Chapter 5, Principled Approaches for Bayesian Deep Learning, introduces well-principled methods for Bayesian neural network approximation.

Chapter 6, Using the Standard Toolbox for Bayesian Deep Learning, introduces approaches for facilitating model uncertainty estimation with common deep learning methods.

Chapter 7, Practical Considerations for Bayesian Deep Learning, explores and compares the advantages and disadvantages of the methods introduced in *Chapter 5* and *Chapter 6*.

Chapter 8, Applying Bayesian Deep Learning, gives a practical overview of a variety of applications of Bayesian Deep Learning, such as detecting out-of-distribution data or robustness against dataset shift.

Chapter 9, Next Steps in Bayesian Deep Learning, discusses some of the latest trends in Bayesian deep learning.

To get the most out of this book

You are expected to have some prior knowledge of machine learning and deep learning, as well as some familiarity with concepts around Bayesian inference. Some practical knowledge of working with Python and a machine learning framework such as TensorFlow or PyTorch would also be valuable but is not necessary.

Python 3.8 or above is recommended, as all code has been tested with Python 3.8. *Chapter 1* provides detailed instructions on setting up your environment for the book's code examples.

Download the example code files

The code bundle for the book is also hosted on GitHub at `https://github.com/PacktPu blishing/Enhancing-Deep-Learning-with-Bayesian-Inference`. If there is an update to the code, it will be updated on the existing GitHub repository.

We also have other code bundles from our rich catalog of books and videos available at `https://github.com/PacktPublishing/`. Check them out!

Download the color images

We also provide a PDF file that has color images of the screenshots/diagrams used in this book. You can download it here: `https://packt.link/7xy10`.

Conventions used

There are a number of text conventions used throughout this book.

`CodeInText`: Indicates code words in text, database table names, folder names, filenames, file extensions, pathnames, dummy URLs, and user input. Here is an example: "Any attempt to run code that has such issues will immediately cause the interpreter to fail, raising a `SyntaxError` exception."

A block of code is set as follows:

```
1  {const set = function(...items) {
2      this.arr  = [...items];
3      this.add = {function}(item) {
4          if( this._arr.includes(item) ) {
5              return false; (SC-Source)}
```

Any command-line input or output is written as follows:

```
$ python3 script.py
```

Some code examples will represent the input of shells. You can recognize them by specific prompt characters:

- »> for interactive Python shell
- $ for Bash shell (macOS and Linux)
- > for CMD or PowerShell (Windows)

Warnings or important notes appear like this.

Important note

Warnings or important notes appear like this.

Tips and tricks appear like this.

Tips or tricks

Appear like this.

Get in touch

Feedback from our readers is always welcome.

General feedback: If you have questions about any aspect of this book, mention the book title in the subject of your message and email us at customercare@packtpub.com.

Errata: Although we have taken every care to ensure the accuracy of our content, mistakes do happen. If you have found a mistake in this book, we would be grateful if you would report this to us. Please visit www.packtpub.com/submit-errata, selecting your book, clicking on the Errata Submission Form link, and entering the details.

Piracy: If you come across any illegal copies of our works in any form on the Internet, we would be grateful if you would provide us with the location address or website name. Please contact us at copyright@packtpub.com with a link to the material.

If you are interested in becoming an author: If there is a topic that you have expertise in and you are interested in either writing or contributing to a book, please visit authors.packtpub.com.

Share Your Thoughts

Once you've read **Enhancing Deep Learning with Bayesian Inference**, we'd love to hear your thoughts! Scan the QR code below to go straight to the Amazon review page for this book and share your feedback.

https://packt.link/r/1-803-24688-X

Your review is important to us and the tech community and will help us make sure we're delivering excellent quality content.

Download a free PDF copy of this book

Thanks for purchasing this book!

Do you like to read on the go but are unable to carry your print books everywhere?

Is your eBook purchase not compatible with the device of your choice?

Don't worry, now with every Packt book you get a DRM-free PDF version of that book at no cost.

Read anywhere, any place, on any device. Search, copy, and paste code from your favorite technical books directly into your application.

The perks don't stop there, you can get exclusive access to discounts, newsletters, and great free content in your inbox daily

Follow these simple steps to get the benefits:

1. Scan the QR code or visit the link below

https://packt.link/free-ebook/9781803246888

2. Submit your proof of purchase

3. That's it! We'll send your free PDF and other benefits to your email directly

1

Bayesian Inference in the Age of Deep Learning

Over the last fifteen years, **machine learning (ML)** has gone from a relatively little-known field to a buzzword in the tech community. This is due in no small part to the impressive feats of **neural networks (NNs)**. Once a niche underdog in the field, **deep learning**'s accomplishments in almost every conceivable application have resulted in a near-meteoric rise in its popularity. Its success has been so pervasive that, rather than being impressed by features afforded by deep learning, we've come to *expect* them. From applying filters in social networking apps, through to relying on Google Translate when on vacation abroad, it's undeniable that deep learning is now well and truly embedded in the technology landscape.

But, despite all of its impressive accomplishments, and the variety of products and features it's afforded us, deep learning has not yet surmounted its final hurdle. As sophisticated neural networks are increasingly applied in mission-critical and safety-critical applications, the questions around their robustness become more and more pertinent. The black-box nature of many deep learning algorithms makes them daunting candidates for safety-savvy solutions architects - so much so that many would prefer sub-standard performance over the potential risks of an opaque system.

So, how can we conquer the apprehension surrounding deep learning and ensure that we create more robust, trustworthy models? While some of the answers to this lie down the path of **explainable artificial intelligence (XAI)**, an important building block lies in the field of **Bayesian deep learning (BDL)**. Through this book, you will discover the fundamental principles behind BDL through practical examples, allowing you to develop a strong understanding of the field, and equipping you with the knowledge and tools you need to build your own BDL models.

But, before we get started, let's delve deeper into the justifications of BDL, and why typical deep learning methods may not be as robust as we'd like. In this chapter, we'll learn about some of the key successes and failures of deep learning, and how BDL can help us to avoid the potentially tragic consequences of standard deep models. We'll then outline the core topics of the rest of the book, before introducing you to the libraries and data that we'll be using in practical examples.

These topics will be covered in the following sections:

- Wonders of the deep learning age

- Understanding the limitations of deep learning

- Core topics

- Setting up the work environment

1.1 Technical requirements

All of the code for this book can be found on the GitHub repository for the book: `https://github.com/PacktPublishing/Enhancing-Deep-Learning-with-Bayesian-Inference`.

1.2 Wonders of the deep learning age

Over the last 10 to 15 years, we've seen a dramatic shift in the landscape of ML thanks to the enormous success of deep learning. Perhaps one of the most impressive feats of the universal impact of deep learning is that it has affected fields from medical imaging and manufacturing all the way through to tools for translation and content creation.

While deep learning has only seen great success over recent years, many of its core principles are already well established. Researchers have been working with neural networks for some time – in fact, one could argue that the first neural network was introduced by Frank Rosenblatt as early as 1957! This, of course, wasn't as sophisticated as the models we have today, but it was an important component of these models: the perceptron, as shown in *Figure 1.1*.

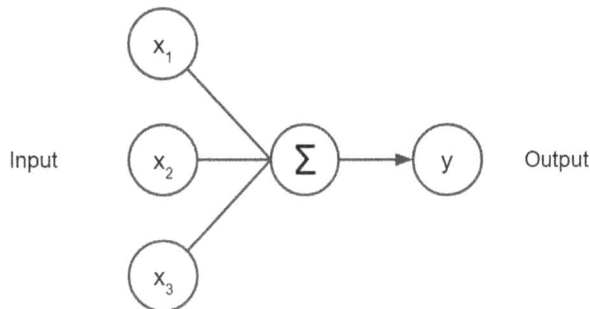

Figure 1.1: Diagram of a single perceptron

The 1980s saw the introduction of many now-familiar concepts, with the introduction of **convolutional neural networks (CNNs)** by Kunihiko Fukushima in 1980, and the development of the **recurrent neural network (RNN)** by John Hopfield in 1982. The 1980s and 1990s saw further maturation of these technologies: Yann LeCun famously applied back-propagation to create a CNN capable of recognizing hand-written digits in 1989, and the crucial concept of long short-term memory RNNs was introduced by Hochreiter and Schmidhuber in 1997.

But, while we had the foundation of today's powerful models before the turn of the century, it wasn't until the introduction of modern GPUs that the field really took off. With the introduction of accelerated training and inference afforded by GPUs, it became possible to develop networks with dozens (or even hundreds) of layers. This opened the door to incredibly sophisticated neural network architectures capable of learning compact feature representations of complex, high-dimensional data.

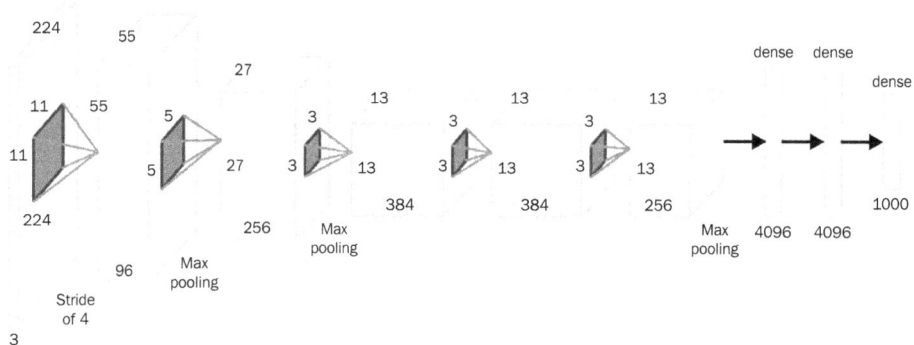

Figure 1.2: Diagram of AlexNet

One of the first highly influential network architectures was AlexNet. This network, developed by Alex Krizhevsky, Ilya Sutskever, and Geoffrey Hinton, comprised 11 layers and was capable of classifying images into one of 1,000 possible classes. It achieved unprecedented performance on the ImageNet Large Scale Visual Recognition Challenge in 2012, illustrating the power of deep networks. AlexNet was the first of an array of influential neural network architectures, and the following years saw the introduction of many now-familiar architectures, including VGG Net, the Inception architectures, ResNet, EfficientNet, YOLO... The list goes on!

But NNs weren't just successful in computer vision applications. In 2014, work by Dzmitry Bahdanau, Kyunghyun Cho, and Yoshua Bengio demonstrated that end-to-end NN models could be used to obtain state-of-the-art results in machine translation. This was a watershed moment for the field, and large-scale machine translation services quickly adopted these end-to-end networks, spurring further advancements in natural language processing. Fast-forward to today, and these concepts have matured to produce the **transformer** architecture – an architecture that has had a dramatic effect on deep learning through its ability to learn rich feature embeddings through self-supervised learning.

With the impressive flexibility granted to them by the wide variety of architectures, neural networks have now achieved state-of-the-art performance in applications across almost every conceivable field, and they're now a familiar part of our daily lives. Whether it's the facial recognition we use on our mobile devices, translation services such as Google Translate, or speech recognition in our smart devices, it's clear that these networks are not just competitive in image classification challenges, they're now an important part of the technologies we're developing, and they're even capable of *outperforming humans.*

While reports of deep learning models outperforming human experts are becoming more and more frequent, the most profound examples are perhaps those in medical imaging. In 2020, a network developed by researchers at Imperial College London and Google Health outperformed six radiologists when detecting breast cancer from mammograms. A few months later, a study from February 2021 demonstrated that a deep learning model was able to outperform two human experts in diagnosing gallbladder disorders. Another study published later that year showed that a CNN outperformed 157 dermatologists in detecting melanoma from images of skin abnormalities.

All of the applications we've discussed so far have been supervised applications of ML, in which models have been trained for classification or regression problems. However, some of the most impressive feats of deep learning are found in other applications, including generative modeling and reinforcement learning. Perhaps one of the most famous examples of the latter is **AlphaGo**, a reinforcement learning model developed by DeepMind. The algorithm, as indicated in its name, was trained to play the game Go via reinforcement learning. Unlike some games, such as chess, which can be solved via fairly straightforward artificial intelligence methods, Go is far more challenging from a computational standpoint. This is due to the sophisticated nature of the game – the many possible combinations of moves are difficult for more traditional approaches. Thus, when

AlphaGo successfully beat Go champions Fan Hui and Lee Sedol, in 2015 and 2016 respectively, this was big news.

DeepMind went on to further refine AlphaGo by creating a version which learned by playing games against itself – AlphaGo Zero. This model was superior to any previous model, achieving superhuman performance in Go. The algorithm at the core of its success, AlphaZero, went on to achieve superhuman performance in a range of other games, proving the algorithm's ability to generalize to other applications.

Another significant milestone for deep learning over the last decade was the introduction of **Generative Adversarial Networks**, or **GANs**. GANs work by employing two networks. The goal of the first network is to generate data with the same statistical qualities as a training set. The goal of the second network is to classify the output of the first network, using what it has learned from the dataset. Because the first network is not trained directly on the data, it does not learn to simply replicate data – instead, it effectively learns to deceive the second network. This is why the term *adversarial* is used. Through this process, the first network is able to learn which kinds of outputs successfully deceive the second network, and thus is able to generate content that matches the data distribution.

GANs can produce particularly impressive outputs. For example, the following image was generated by the StyleGAN2 model:

Figure 1.3: Face generated by StyleGAN2 from thispersondoesnotexist.com.

But GANs aren't just useful for generating realistic faces; they have practical applications in many other fields, such as suggesting molecular combinations for drug discovery. They are also a powerful tool for improving other ML methods through data augmentation – using the GAN-generated data to augment datasets.

All these successes may make deep learning seem infallible. While its achievements are impressive, they don't tell the whole story. In the next section, we'll learn about some of deep learning's failings, and start to understand how Bayesian approaches may help to avoid these in the future.

1.3 Understanding the limitations of deep learning

As we've seen, deep learning has achieved some remarkable feats, and it's undeniable that it's revolutionizing the way that we deal with data and predictive modeling. But deep learning's short history also comprises darker tales: stories that bring with them crucial lessons for developing systems that are more robust, and, crucially, safer.

In this section, we'll introduce a couple of key cases in which deep learning failed, and we will discuss how a Bayesian perspective could have helped to produce a better outcome.

1.3.1   Bias in deep learning systems

We'll start with a textbook example of **bias**, a crucial problem faced by data-driven methods. This example centers around Amazon. Now a household name, the e-commerce company started out by revolutionizing the world of book retail, before becoming literally *the* one-stop shop for just about anything: from garden furniture to a new laptop, or even a home security system, if you can imagine it, you can probably purchase it on Amazon. The company has also been responsible for significant strides technologically, often as a means of improving its infrastructure in order to enable its expansion. From hardware infrastructure to theoretical and technological leaps in optimization methods, what started out as an e-commerce organization has now become one of the key figures in technology.

While these technological leaps often set the standard for the industry, this example did the opposite: demonstrating a key weakness of data-driven methods. The case we're referring to is that of Amazon's AI recruiting software. With automation playing such a key role in so much of Amazon's success, it made sense to expand this automation to reviewing resumes. In 2014, Amazon's ML engineers deployed a tool to do just that. Trained on the previous 10 year's worth of applicants, the tool was designed to learn to identify favorable traits from the company's enormous pool of applicants. However, in 2015 it became clear that it had latched onto certain features that resulted in deeply undesirable behavior.

The issue was largely due to the underlying data: because of the nature of the tech industry at the time, Amazon's dataset of resumes was dominated by male applicants. This resulted in tremendous inequity in the model's predictions: it effectively learned to favor men, becoming hugely biased against female applicants. The discriminatory behavior of the model resulted in the project being abandoned by Amazon, and it now serves as a key example of bias for the AI community.

An important factor to consider in the problem presented here is that this bias isn't just driven by *explicit* information, such as a person's name (which could be a clue as to their gender): algorithms learn latent information, which can then drive bias. This means the problem can't simply be solved by anonymizing people – it's up to the engineers and scientists to ensure that bias is evaluated comprehensively so that the algorithms we deploy are fair. While Bayesian methods can't make bias disappear, they present us with a range of tools that can help with these problems. As we'll see later in the book, Bayesian methods give us the ability to determine whether data is in-distribution or **out-of-distribution (OOD)**. In this case, Amazon could have used this capability of Bayesian methods: separating the OOD data and analyzing it to understand why it was OOD. Was it picking up on things that were relevant, such as applicants with the wrong kind of experience? Or was it picking up on something irrelevant and discriminatory, such as the applicant's gender? This could have helped Amazon's ML team to spot the undesirable behavior early, allowing them to develop an unbiased solution.

1.3.2   The danger of over-confident predictions

Another widely referenced example of a deep learning failure is illustrated in the paper *Robust Physical-World Attacks on Deep Learning Visual Classification* by Kevin Eykholt *et al.* (https://arxiv.org/abs/1707.08945). This paper played an important role in highlighting the issue of **adversarial attacks** on deep learning models: slightly modifying input data so that the model produces an incorrect prediction. In one of the key examples from their paper, they stick white and black stickers to a stop sign. While the modifications to the sign were subtle, the computer vision model interpreted the modified sign as a Speed Limit 45 sign.

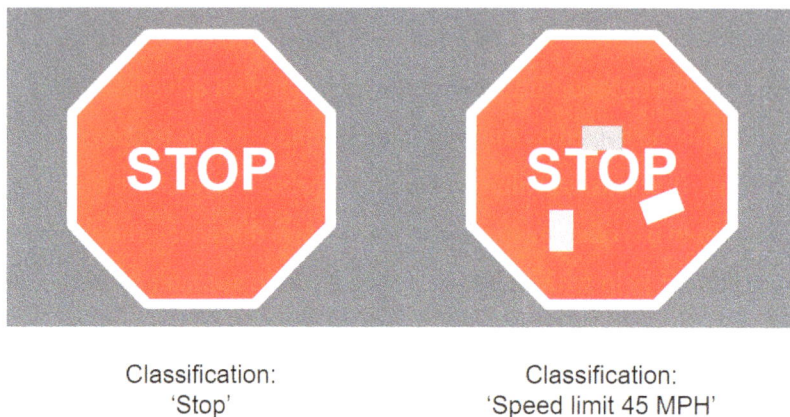

Classification: Classification:
'Stop'   'Speed limit 45 MPH'

Figure 1.4: Illustration of the effect of a simple adversarial attack on a model interpreting a stop sign.

At first, this may seem inconsequential, but if we take a step back and consider the amount of work that Tesla, Uber, and others have dedicated towards self-driving cars, it's easy to see how this sort of adversarial perturbation could lead to catastrophic consequences. In the case of this sign, this misclassification could lead to a self-driving car bypassing a stop sign, hurtling into traffic at an intersection. This would obviously not be good for the passengers or other road users. In fact, an incident not too dissimilar to what we're describing here happened in 2016 when a Tesla Model S collided with a truck in northern Florida (`https://www.reuters.com/article/us-tesla-crash-idUSKBN19A2XC`). According to Tesla, the trailer being pulled by the truck wasn't detected by Tesla's autopilot as it couldn't distinguish it from the backdrop of the bright sky behind the trailer. The driver also didn't notice the trailer, ultimately resulting in a fatal collision. But what if the decision processes used by the autopilot were more sophisticated? One of the key themes throughout this book is that of making *robust* decisions with our ML systems, particularly in mission-critical or safety-critical applications.

While this traffic sign example provides an intuitive illustration of the dangers associated with misclassifications, this applies to a vast range of other scenarios, from robotics equipment used for manufacturing through to automated surgical procedures.

Having some idea of confidence (or uncertainty) is an important step towards improving the robustness of these systems and ensuring consistently safe behavior. In the case of the stop sign, simply having a model that "knows when it doesn't know" can prevent potentially tragic outcomes. As we'll see later in the book, BDL methods allow us to detect adversarial inputs through their uncertainty estimates. In our self-driving car example, this could be incorporated in the logic so that, if the model is uncertain, the car safely comes to a stop, switching to manual mode to allow the driver to safely navigate the situation. This is the *wisdom* that comes with uncertainty-aware models: allowing us to design models that know their limitations, and thus are more robust in unexpected scenarios.

1.3.3   Shifting trends

Our last examples look at the challenge of dealing with data that changes over time – a common problem in real-world applications. The first problem we'll consider, typically referred to as **dataset shift** or **covariate shift**, occurs when the data encountered by a model at inference time changes relative to the data the model was trained on. This is often due to the dynamic nature of real-world problems and the fact that training sets – even very large training sets – rarely represent the total variation present in the phenomena they represent. An important example of this can be found in the paper *Systematic Review of Approaches to Preserve Machine Learning Performance in the Presence of Temporal Dataset Shift in Clinical Medicine*, in which Lin Lawrence Guo *et al.* highlight concerns around dataset shift (https://www.ncbi.nlm.nih.gov/pmc/articles/PMC8410238/). Their work shows that there is relatively little literature on tackling issues related to dataset shift in ML models applied in clinical settings. This is problematic because clinical data is dynamic. Let's consider an example.

In our example, we have a model that's trained to automatically prescribe medication for a patient given their symptoms. A patient complains to a physician about respiratory symptoms, and the physician uses the model to prescribe medication. Because of the data presented to the model, it prescribes antibiotics. This works for many patients for a while, but over time something changes: a new disease becomes prevalent in the population. The new disease happens to have very similar symptoms to the bacterial infection that was going around previously, but this is caused by a virus. Because the model isn't capable of adapting to dataset shift, it continues recommending antibiotics. Not only will they not help the patients, but it could contribute to antibiotic resistance within the local population.

In order to be robust to these shifts in real-world data, models need to be sensitive to dataset shift. One way to do this is through the use of Bayesian methods, which provide uncertainty estimates. Applying this to our automatic prescriber example, the model becomes sensitive to small changes in the data when capable of producing uncertainty estimates. For example, there may be subtle differences in symptoms, such as a different type of cough, associated with our new viral infection. This will cause the uncertainty associated with the model predictions to rise, indicating that the model needs to be updated with new data.

A related issue, referred to as **catastrophic forgetting**, is caused by models adapting to changes in data. Given our example, this sounds like a good thing: if models are adapting to changes in the data, then they're always up to date, right? Unfortunately, it's not quite so simple. Catastrophic forgetting occurs when models learn from new data, but "forget" about past data in the process.

For example, say an ML algorithm is developed to identify fraudulent documents. It may work very well at first, but fraudsters will quickly notice that methods that used to fool automated document verification no longer work, so they develop new methods. While a few of these methods get through, the model – using its uncertainty estimates – notices that it needs to adapt to the new data. The model updates its dataset, focusing on the current popular attack methods, and runs a few more training iterations. Once again, it successfully thwarts the fraudsters, but, much to the surprise of the model's designers, the model has started letting through older, less sophisticated attacks: attacks that used to be easy for it to identify.

In training on the new data, the model's parameters have changed. Because there wasn't sufficient support for the old data in the updated dataset, the model has lost information about old associations between the inputs (documents) and their classification (whether or not they're fraudulent).

While this example used uncertainty estimates to tackle the issue of dataset shift, it could have further leveraged them to ensure that its dataset was balanced. This can be done using methods such as **uncertainty sampling**, which look to sample from uncertain regions, ensuring that the dataset used to train the model captures all available information from current and past data.

1.4   Core topics

The aim of this book is to provide you with the tools and knowledge you need to develop your own BDL solutions. To this end, while we assume some familiarity with concepts of statistical learning and deep learning, we will still provide a refresher of these fundamental concepts.

In *Chapter 2*, we'll go over some of the key concepts from Bayesian inference, including probabilities and model uncertainty estimates. In *Chapter 3*, we'll cover important key aspects of deep learning, including learning via backpropagation, and popular varieties of NNs. With these fundamentals covered, we'll start to explore BDL in *Chapter 4*. In *Chapters 5* and *6* we'll delve deeper into BDL; we'll first learn about principled methods, before going on to understand more practical methods for approximating Bayesian neural networks.

In *Chapter 7*, we'll explore some practical considerations for BDL, helping us to understand how best to apply these methods to real-world problems. By *Chapter 8*, we should have a strong understanding of the core BDL methods, and we'll cement this with a number of practical examples. Finally, *Chapter 9* will provide an overview of the current challenges within the field of BDL and give you an idea of where the technology is headed.

Throughout most of the book, the theory will be accompanied by hands-on examples, allowing you to develop a strong understanding by implementing these methods yourself. In order to follow these coding examples, you will need to have a Python environment set up with the necessary prerequisites. We'll go over these in the next section.

1.5   Setting up the work environment

To complete the practical elements of the book, you'll need a Python 3.9 environment with the necessary prerequisites. We recommend using conda, a Python package manager specifically designed for scientific computing applications. To install conda, simply head to https://conda.io/projects/conda/en/latest/user-guide/install/index.html and follow the instructions for your operating system.

With conda installed, you can set up the conda environment that you'll use for the book:

```
1   conda create -n bdl python=3.9
```

When you hit *Enter* to execute this command, you'll be asked if you wish to continue installing the required packages; simply type y and hit **Enter**. conda will now proceed to install the core packages.

You can now activate your environment by typing the following:

```
1   conda activate bdl
```

You'll now see that your shell prompt contains bdl, indicating that your conda environment is active. Now you're ready to install the prerequisites for the book. The key libraries required for the book are as follows:

- **NumPy**: Numerical Python, or NumPy, is the core package for numerical programming in Python. You're likely very familiar with this already.

- **SciPy**: SciPy, or Scientific Python, provides the fundamental packages for scientific computing applications. The full scientific computing stack comprising SciPy, matplotlib, NumPy, and other libraries, is often referred to as the SciPy stack.

- **scikit-learn**: This is the core Python machine learning library. Built on the SciPy stack, it provides easy-to-use implementations of many popular ML methods. It also provides a substantial number of helper classes and functions for data loading and processing, which we'll use throughout the book.

- **TensorFlow**: TensorFlow, along with PyTorch and JAX, is one of the popular Python deep learning frameworks. It provides the tools necessary for developing deep learning models, and it will provide the foundation for many of the programming examples throughout the book.

- **TensorFlow Probability**: Built on TensorFlow, TensorFlow Probability provides the tools necessary for working with probabilistic neural networks. We'll be using this along with TensorFlow for many of the Bayesian neural network examples.

To install the full list of dependencies required for the book, with your conda environment activated, enter the following:

```
1  conda install -c conda-forge scipy sklearn matplotlib seaborn
2  tensorflow tensorflow-probability
```

Let's summarize what we have learned.

1.6 Summary

In this chapter, we've revisited the successes of deep learning, renewing our understanding of its enormous potential, and its ubiquity within today's technology. We've also explored some key examples of its shortcomings: scenarios in which deep learning has failed us, demonstrating the potential for catastrophic consequences. While BDL can't eliminate these risks, it can allow us to build more robust ML systems that incorporate both the flexibility of deep learning and the caution of Bayesian inference.

In the next chapter, we'll dive deeper into the latter as we cover some of the core concepts of Bayesian inference and probability, in preparation for our foray into BDL.

2

Fundamentals of Bayesian Inference

Before we get into Bayesian inference with **Deep Neural Networks (DNNs)**, we should take some time to understand the fundamentals. In this chapter, we'll do just that: exploring the core concepts of Bayesian modeling, and taking a look at some of the popular methods used for Bayesian inference. By the end of this chapter, you should have a good understanding of why we use probabilistic modeling, and what kinds of properties we look for in well principled – or well conditioned – methods.

This content will be covered in the following sections:

- Refreshing our knowledge of Bayesian modeling

- Bayesian inference via sampling

- Exploring the Gaussian processes

2.1 Refreshing our knowledge of Bayesian modeling

Bayesian modeling is concerned with understanding the probability of an event occurring given some prior assumptions and some observations. The prior assumptions describe our initial beliefs, or hypothesis, about the event. For example, let's say we have two six-sided dice, and we want to predict the probability that the sum of the two dice is 5. First, we need to understand how many possible outcomes there are. Because each die has 6 sides, the number of possible outcomes is $6 \times 6 = 36$. To work out the possibility of rolling a 5, we need to work out how many combinations of values will sum to 5:

	1	2	3	4	5	6
1	2	3	4	5	6	7
2	3	4	5	6	7	8
3	4	5	6	7	8	9
4	5	6	7	8	9	10
5	6	7	8	9	10	11
6	7	8	9	10	11	12

Figure 2.1: Illustration of all values summing to five when rolling two six-sided dice

As we can see here, there are 4 combinations that add up to 5, thus the probability of having two dice produce a sum of 5 is $\frac{4}{36}$, or $\frac{1}{9}$. We call this initial belief the **prior**. Now, what happens if we incorporate information from an observation? Let's say we know what the value for one of the dice will be – let's say 3. This shrinks our number of possible values down to 6, as we only have the remaining die to roll, and for the result to be 5, we'd need this value to be 2.

	1	2	3	4	5	6
1	2	3	4	5	6	7
2	3	4	5	6	7	8
3	4	5	6	7	8	9
4	5	6	7	8	9	10
5	6	7	8	9	10	11
6	7	8	9	10	11	12

Figure 2.2: Illustration of remaining value, which sums to five after rolling the first die

Because we assume our die is fair, the probability of the sum of the dice being 5 is now $\frac{1}{6}$. This probability, called the **posterior**, is obtained using information from our observation. At the core of Bayesian statistics is Bayes' rule (hence "Bayesian"), which we use to determine the posterior probability given some prior knowledge. Bayes' rule is defined as:

$$P(A|B) = \frac{P(B|A) \times P(A)}{P(B)} \tag{2.1}$$

Where we can define $P(A|B)$ as $P(d_1 + d_2 = 5|\mathbf{d_1 = 3})$, where d_1 and d_2 represent dice 1 and 2 respectively. We can see this in action using our previous example. Starting with the **likelihood**, that is, the term on the left of our numerator, we see that:

$$P(B|A) = P(d_1 = 3|d_1 + d_2 = 5) = \frac{1}{4} \tag{2.2}$$

We can verify this by looking at our grid. Moving to the second part of the numerator – the prior – we see that:

$$P(A) = P(d_1 + d_2 = 5) = \frac{4}{36} = \frac{1}{9} \tag{2.3}$$

On the denominator, we have our **normalization constant** (also referred to as the **marginal likelihood**), which is simply:

$$P(B) = P(d_1 = 3) = \frac{1}{6} \tag{2.4}$$

Putting this all together using Bayes' theorem, we have:

$$P(d_1 + d_2 = 5|d_1 = 3) = \frac{\frac{1}{4} \times \frac{1}{9}}{\frac{1}{6}} = \frac{1}{6} \tag{2.5}$$

What we have here is the *probability* of the outcome being 5 if we know one die's value. However, in this book, we'll often be referring to **uncertainties** rather than probabilities – and learning methods to obtain uncertainty estimates with DNNs. These methods belong to a broader class of **uncertainty quantification**, and aim to quantify the uncertainty in the predictions from an ML model. That is, we want to predict $P(\hat{y}|\theta)$, where $\hat{y}$ is a prediction from a model, and θ represents the parameters of the model.

As we know from fundamental probability theory, probabilities are bound between 0 and 1. The closer we are to 1, the more likely – or probable – the event is. We can view our uncertainty as subtracting our probability from 1. In the context of the example here, the probability of the sum being 5 is $P(d_1 + d_2 = 5|d_1 = 3) = \frac{1}{6} = 0.166$. So, our uncertainty is simply $1 - \frac{1}{6} = \frac{5}{6} = 0.833$, meaning that there's a $> 80\%$ chance that the outcome *will not* be 5. As we proceed through the book, we'll learn about different sources of uncertainty, and how uncertainties can help us to develop more robust deep learning systems.

Let's continue using our dice example to build a better understanding of for model uncertainty estimates. Many common machine learning models work on the basis of **maximum likelihood estimation** or **MLE**. That is, they look to predict the value that is *most likely*: tuning their parameters during training to produce the most likely outcome $\hat{y}$ given some input x. As a simple illustration, let's say we want to predict the value of $d_1 + d_2$ given a value of d_1. We can simply define this as the **expectation** of $d_1 + d_2$ conditioned on d_1:

$$\hat{y} = \mathbb{E}[d_1 + d_2 | d_1] \tag{2.6}$$

That is, the *mean* of the possible values of $d_1 + d_2$.

Setting $d_1 = 3$, our possible values for $d_1 + d_2$ are $\{4, 5, 6, 7, 8, 9\}$ (as illustrated in *Figure 2.2*), making our mean:

$$\mu = \frac{1}{6} \sum_{i=1}^{6} a_i = \frac{4 + 5 + 6 + 7 + 8 + 9}{6} = 6.5 \tag{2.7}$$

This is the value we'd get from a simple linear model, such as a linear regression defined by:

$$\hat{y} = \beta x + \xi \tag{2.8}$$

In this case, the values of our intersection and bias are $\beta = 1, \xi = 3.5$. If we change our value of d_1 to 1, we see that this mean changes to 4.5 – the mean of the set of possible values of $d_1 + d_2 | d_1 = 1$, in other words $\{2, 3, 4, 5, 6, 7\}$. This perspective on our model predictions is important: while this example is very straightforward, the same principle applies to far more sophisticated models and data. The value we typically see with ML models is the *expectation*, otherwise known as the mean. As you are likely aware, the mean is often referred to as the **first statistical moment** – with the **second statistical moment** being the **variance**, and the variance allows us to quantify uncertainty.

The variance for our simple example is defined as follows:

$$\sigma^2 = \frac{\sum_{i=1}^{6}(a_i - \mu)^2}{n - 1} \tag{2.9}$$

These statistical moments should be familiar to you, as should the fact that the variance here is represented as the square of the **standard deviation**, σ. For our example here, for which we assume d_2 is a fair die, the variance will always be constant: $\sigma^2 = 2.917$. That is to say, given any value of d_1, we know that values of d_2 are all equally likely, so the uncertainty does not change. But what if we have an unfair die d_2, which has a 50% chance of landing on a 6, and a 10% chance of landing on each other number? This changes both our mean and our variance. We can see this by looking at how we would represent this as a set of possible values (in other words, a perfect sample of the die) – the set of possible values for $d_1 + d_2 | d_1 = 1$ now becomes $\{2, 3, 4, 5, 6, 7, 7, 7, 7, 7\}$. Our new model will now have a bias of $\xi = 4.5$, making our prediction:

$$\hat{y} = 1 \times 1 + 4.5 = 5.5 \tag{2.10}$$

We see that the expectation has increased due to the change in the underlying probability of the values of die d_1. However, the important difference here is in the change in the variance value:

$$\sigma^2 = \frac{\sum_{i=1}^{10}(a_i - \mu)^2}{n - 1} = 3.25 \tag{2.11}$$

Our variance has *increased*. As variance essentially gives us the average of the distance of each possible value from the mean, this shouldn't be surprising: given the weighted die, it's more likely that the outcome will be distant from the mean than with an unweighted die, and thus our variance increases. To summarize, in terms of uncertainty: the greater the likelihood that the outcome will be further from the mean, the greater the uncertainty.

This has important implications for how we interpret predictions from machine learning models (and statistical models more generally). If our predictions are an approximation of the mean, and our uncertainty quantifies how likely it is for an outcome to be distant from the mean, then our uncertainty tells us **how likely it is that our model prediction is incorrect**. Thus, model uncertainties allow us to decide when to trust the predictions, and when we should be more cautious.

The examples given here are very basic, but should help to give you an idea of what we're looking to achieve with model uncertainty quantification. We will continue to explore these concepts as we learn about some of the benchmark methods for Bayesian inference, learning how these concepts apply to more complex, real-world problems. We'll start with perhaps the most fundamental method of Bayesian inference: sampling.

2.2   Bayesian inference via sampling

In practical applications, it's not possible to know exactly what a given outcome would be, and, similarly, it's not possible to observe all possible outcomes. In these cases, we need to make a best estimate based on the evidence we have. The evidence is formed of **samples** – observations of possible outcomes. The aim of ML, broadly speaking, is to learn models that generalize well from a subset of data. The aim of Bayesian ML is to do so while also providing an estimate of the uncertainty associated with the model's predictions. In this section, we'll learn about how we can use sampling to do this, and will also learn why sampling may not be the most sensible approach.

2.2.1   Approximating distributions

At the most fundamental level, sampling is about approximating distributions. Say we want to know the distribution of the height of people in New York. We could go out and measure everyone, but that would involve measuring the height of 8.4 million people! While this would give us our most accurate answer, it's also a deeply impractical approach.

Instead, we can sample from the population. This gives us a basic example of **Monte Carlo sampling**, where we use random sampling to provide data from which we can approximate a distribution. For example, given a database of New York residents, we could select – at random – a sub-population of residents, and use this to approximate the height distribution of all residents. With random sampling – and any sampling, for that matter – the accuracy of the approximation is dependent on the size of the sub-population. What we're looking to achieve is a **statistically significant** sub-sample, such that we can be confident in our approximation.

To get a better impression of this, we'll simulate the problem by generating 100,000 data points from a truncated normal distribution, to approximate the kind of height distribution we may see for a population of 100,000 people. Say we draw 10 samples, at random, from our population. Here's what our distribution would look like (on the right) compared with the true distribution (on the left):

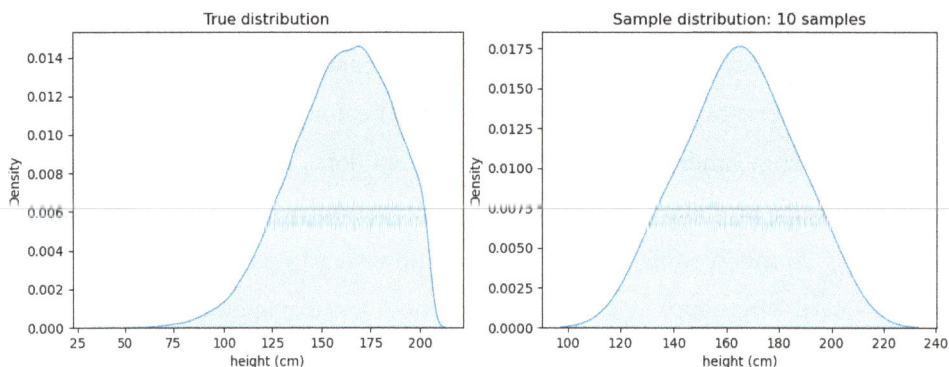

Figure 2.3: Plot of true distribution (left) versus sample distribution (right)

As we can see, this isn't a great representation of the true distribution: what we see here is closer to a triangular distribution than a truncated normal. If we were to infer something about the population's height based on this distribution alone, we'd arrive at a number of inaccurate conclusions, such as missing the truncation above 200 cm, and the tail on the left of the distribution.

We can get a better impression by increasing our sample size – let's try drawing 100 samples:

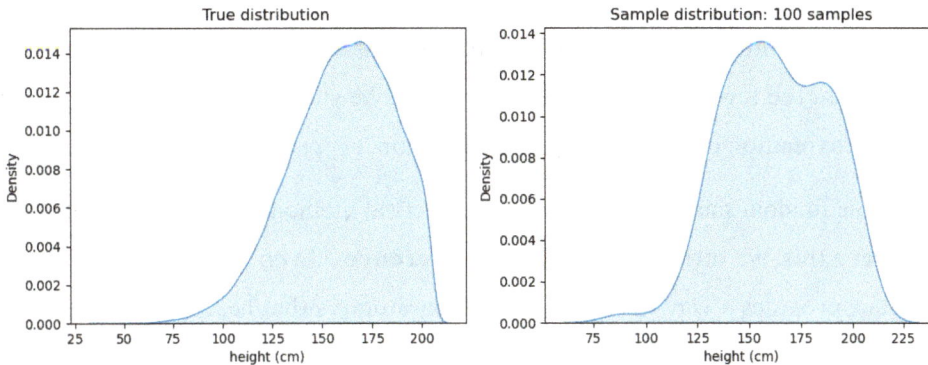

Figure 2.4: Plot of true distribution (left) versus sample distribution (right).

Things are starting to look better: we're starting to see some of the tail on the left as well as the truncation toward 200 cm. However, this sample has sampled more from some regions than others, leading to misrepresentation: our mean has been pulled down, and we're seeing two distinct peaks, rather than the single peak we see in the true distribution. Let's increase our sample size by a further order of magnitude, scaling up to 1,000 samples:

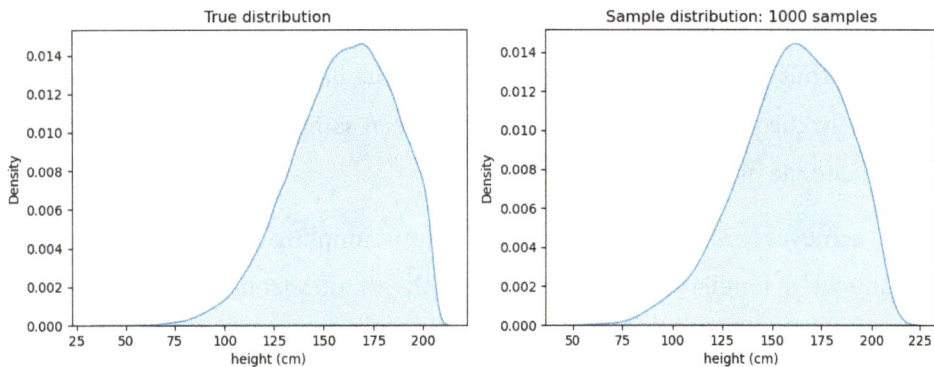

Figure 2.5: Plot of true distribution (left) versus sample distribution (right)

This is looking much better – with a sample set of only one hundredth the size of our true population, we now see a distribution that closely matches our true distribution. This example demonstrates how, through random sampling, we can approximate the true distribution using a significantly smaller pool of observations. But that pool still has to have enough information to allow us to arrive at a good approximation of the true distribution: too few samples and our subset will be statistically *insufficient*, leading to poor approximation of the underlying distribution.

But simple random sampling isn't the most practical method for approximating distributions. To achieve this, we turn to **probabilistic inference**. Given a model, probabilistic inference provides a way to find the model parameters that best describe our data. To do so, we need to first define the type of model – this is our prior. For our example, we'll use a truncated Gaussian: the idea here being, using our intuition, it's reasonable to assume people's height follows a normal distribution, but that very few people are above, say, 6'5." So, we'll specify a truncated Gaussian distribution with an upper limit of 205 cm, or just over 6'5." As it's a Gaussian distribution, in other words, $\mathcal{N}(\mu, \sigma)$, our model parameters are $\theta = \{\mu, \sigma\}$ – with the additional constraint that our distribution has an upper limit of $b = 205$.

This brings us to a fundamental class of algorithms: **Markov Chain Monte Carlo**, or **MCMC** methods. Like simple random sampling, these allow us to build a picture of the true underlying distribution, but they do so sequentially, whereby each sample is dependent on the sample before it. This sequential dependence is known as the **Markov property**, thus the *Markov chain* component of the name. This sequential approach accounts for the probabilistic dependence between samples and allows us to better approximate the probability density.

MCMC achieves this through sequential random sampling. Just as with the random sampling we're familiar with, MCMC randomly samples from our distribution. But, unlike simple random sampling, MCMC considers pairs of samples: some previous sample x_{t-1} and some current sample x_t. For each pair of samples, we have some criteria that specifies whether or not we keep the sample (this varies depending on the particular flavor of MCMC). If the new value meets this criteria, say if x_t is "preferential to" our previous value x_{t-1}, then the sample is added to the chain and becomes x_t for the next round. If the sample doesn't meet the criteria, we stick with the current x_t for the next round. We repeat this over a (usually large) number of iterations, and in the end we should arrive at a good approximation of our distribution.

The result is an efficient sampling method that is able to closely approximate the true parameters of our distribution. Let's see how this applies to our height distribution example. Using MCMC with just 10 samples, we arrive at the following approximation:

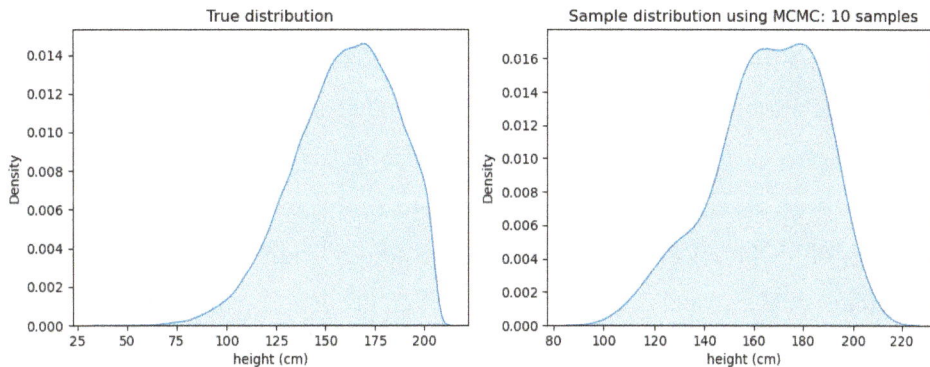

Figure 2.6: Plot of true distribution (left) versus approximate distribution via MCMC (right)

Not bad for ten samples – certainly far better than the triangular distribution we arrived at with simple random sampling. Let's see how we do with 100:

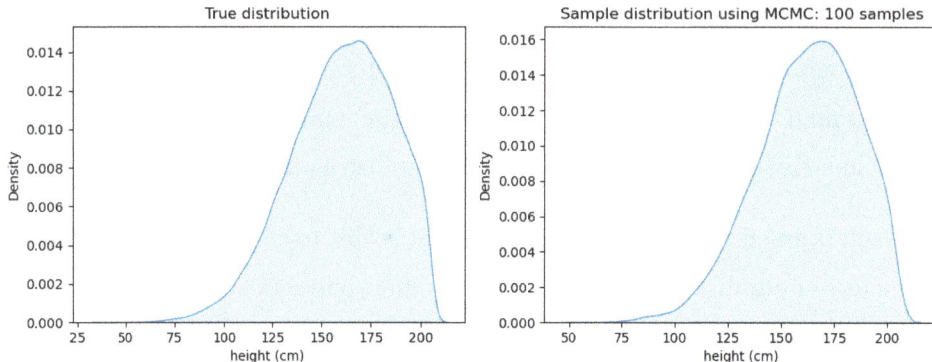

Figure 2.7: Plot of true distribution (left) versus approximate distribution via MCMC (right)

This is looking pretty excellent – in fact, we're able to obtain a better approximation of our distribution with 100 MCMC samples than we are with 1,000 simple random samples. If we continue to larger numbers of samples, we'll arrive at closer and closer approximations of our true distribution. But our simple example doesn't fully capture the power of MCMC: MCMC's true advantage comes from being able to approximate high-dimensional distributions, and has made it an invaluable technique for approximating intractable high-dimensional integrals in a variety of domains.

In this book, we're interested in how we can estimate the probability distribution of the parameters of machine learning models – this allows us to estimate the uncertainty associated with our predictions. In the next section, we'll take a look at how we do this practically by applying sampling to Bayesian linear regression.

2.2.2 Implementing probabilistic inference with Bayesian linear regression

In typical linear regression, we want to predict some output $\hat{y}$ from some input x using a linear function $f(x)$, such that $\hat{y} = \beta x + \xi$. With Bayesian linear regression, we do this probabilistically, introducing another parameter, σ^2, such that our regression equation becomes:

$$\hat{y} = \mathcal{N}(x\beta + \xi, \sigma^2) \tag{2.12}$$

That is, $\hat{y}$ follows a Gaussian distribution.

Here, we see our familiar bias term ξ and intercept β, and introduce a variance parameter σ^2. To fit our model, we need to define a prior over these parameters – just as we did for our MCMC example in the last section. We'll define these priors as:

$$\xi \approx \mathcal{N}(0,1) \tag{2.13}$$

$$\beta \approx \mathcal{N}(0,1) \tag{2.14}$$

$$\sigma^2 \approx |\mathcal{N}(0,1)| \tag{2.15}$$

Note that equation 2.15 denotes the half-normal of a Gaussian distribution (the positive half of a zero-mean Gaussian, as standard deviation cannot be negative). We'll refer to our model parameters as $\theta = \beta, \xi, \sigma^2$, and we'll use sampling to find the parameters that maximise the likelihood of these given our data, in other words, the conditional probability of our parameters given our data D: $P(\theta|D)$.

There are a variety of MCMC sampling approaches we could use to find our model parameters. A common approach is to use the **Metropolis-Hastings** algorithm. Metropolis-Hastings is particularly useful for sampling from intractable distributions. It does so through the use of a proposal distribution, $Q(\theta'|\theta)$, which is proportional to, but not exactly equal to, our true distribution. This means that, for example, if some value x_1 is twice as likely as some other value x_2 in our true distribution, this will be true of our proposal distribution too. Because we're interested in the probability of observations, we don't need to know what the *exact* value would be in our true distribution – we just need to know that, proportionally, our proposal distribution is equivalent to our true distribution.

Here are the key steps of Metropolis-Hastings for our Bayesian linear regression.

First, we initialize with an arbitrary point θ sampled from our parameter space, according to the priors for each of our parameters. Using a Gaussian distribution centered on our first set of parameters θ, select a new point θ'. Then, for each iteration $t \in T$, do the following:

1. Calculate the acceptance criteria, defined as:

$$\alpha = \frac{P(\theta'|D)}{P(\theta|D)} \qquad (2.16)$$

2. Generate a random number from a uniform distribution $\epsilon \in [0, 1]$. If $\epsilon <= \alpha$, accept the new candidate parameters – adding these to the chain, assigning $\theta = \theta'$. If $\epsilon > \alpha$, keep the current θ and draw a new value.

This acceptance criteria means that, if our new set of parameters have a higher likelihood than our last set of parameters, we'll see $\alpha > 1$, in which case $\alpha < \epsilon$. This means that, when we sample parameters that are *more likely* given our data, we'll always accept these parameters. If, on the other hand, $\alpha < 1$, there's a chance we'll reject the parameters, but we may also accept them – allowing us to explore regions of lower likelihood.

These mechanics of Metropolis-Hastings result in samples that can be used to compute high-quality approximations of our posterior distribution. Practically, Metropolis-Hastings (and MCMC methods more generally) requires a burn-in phase – an initial phase of sampling used to escape regions of low density, which are typically encountered given the arbitrary initialization.

Let's apply this to a simple problem: we'll generate some data for the function $y = x^2 + 5 + \eta$, where η is a noise parameter distributed according to $\eta \approx \mathcal{N}(0, 5)$. Using Metropolis-Hastings to fit our Bayesian linear regressor, we get the following fit using the points sampled from our function (represented by the crosses):

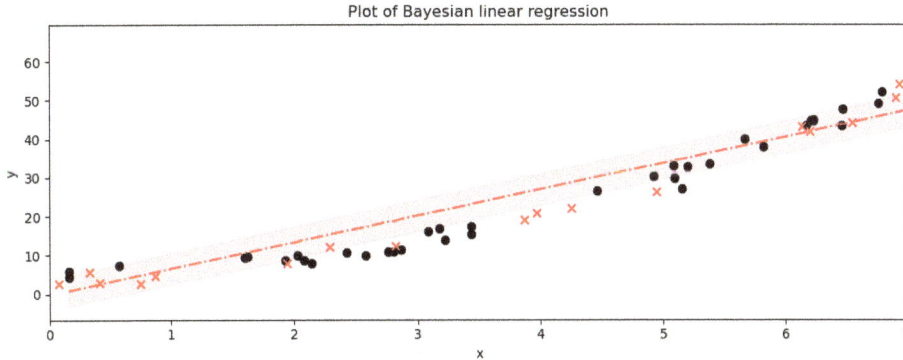

Figure 2.8: Bayesian linear regression on generated data with low variance

We see that our model fits the data in the same way we would expect for standard linear regression. However, unlike standard linear regression, our model produces predictive uncertainty: this is represented by the shaded region. This predictive uncertainty gives an impression of how much our underlying data varies; this makes this model much more useful than a standard linear regression, as now we can get an impression of the spread of our data, as well as the general trend. We can see how this varies if we generate new data and fit again, this time increasing the spread of the data by modifying our noise distribution to $\eta \approx \mathcal{N}(0, 20)$:

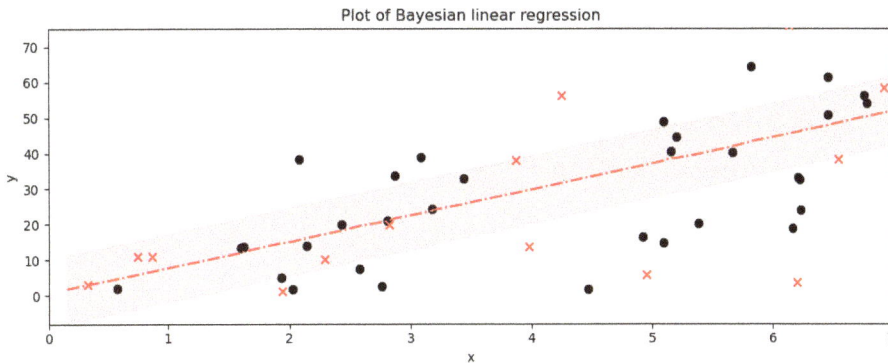

Figure 2.9: Bayesian linear regression on generated data with high variance

We see that our predictive uncertainty has increased proportionally to the spread of the data. This is an important property in uncertainty-aware methods: when we have small uncertainty, we know our prediction fits the data well, whereas when we have large uncertainty, we know to treat our prediction with caution, as it indicates the model isn't fitting this region particularly well. We'll see a better example of this in the next section, which will go on to demonstrate how regions of more or less data contribute to our model uncertainty estimates.

Here, we see that our predictions fit our data pretty well. In addition, we see that σ^2 varies according to the availability of data in different regions. What we're seeing here is a great example of a very important concept, **well calibrated uncertainty** – also termed **high-quality uncertainty**. This refers to the fact that, in regions where our predictions are inaccurate, our uncertainty is also high. Our uncertainty estimates are **poorly calibrated** if we're very confident in regions with inaccurate predictions, or very uncertain in regions with accurate predictions. As it's well-calibrated, sampling is often used as a benchmark for uncertainty quantification.

Unfortunately, while sampling is effective for many applications, the need to obtain many samples for each parameter means that it quickly becomes computationally prohibitive for high dimensions of parameters. For example, if we wanted to start sampling parameters for complex, non-linear relationships (such as sampling the weights of a neural network), sampling would no longer be practical. Despite this, it's still useful in some cases, and later we'll see how various BDL methods make use of sampling.

In the next section, we'll explore the Gaussian process – another fundamental method for Bayesian inference, and a method that does not suffer from the same computational overheads as sampling.

2.3   Exploring the Gaussian process

As we've seen in the previous section, sampling quickly becomes prohibitively expensive. To address this, we can use ML models specifically designed to produce uncertainty estimates – the gold standard of which is the **Gaussian process**.

The Gaussian process, or **GP**, has become a staple probabilistic ML model, seeing use in a broad variety of applications from pharmacology through to robotics. Its success is largely down to its ability to produce high-quality uncertainty estimates over its predictions in a well-principled fashion. So, what do we mean by a Gaussian process?

In essence, a GP is a distribution over functions. To understand what we mean by this, let's take a typical ML use case. We want to learn some function $f(\mathbf{x})$, which maps a series of inputs $\mathbf{x}$ onto a series of outputs $\mathbf{y}$, such that we can approximate our output via $\hat{\mathbf{y}} = f(\mathbf{x})$. Before we see any data, we know nothing about our underlying function; there is an infinite number of possible functions this could be:

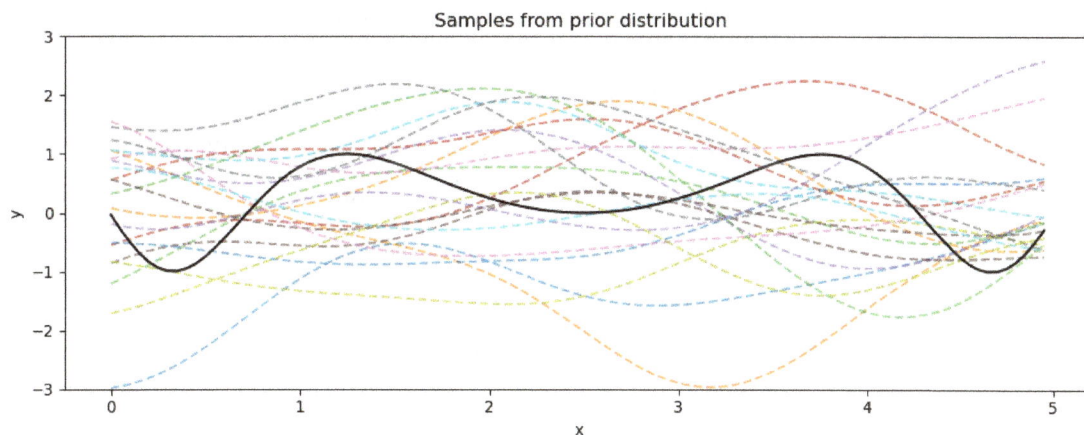

Figure 2.10: Illustration of space of possible functions before seeing data

Here, the black line is the true function we wish to learn, while the dotted lines are the possible functions given the data (in this case, no data). Once we observe some data, we see that the number of possible functions becomes more constrained, as we see here:

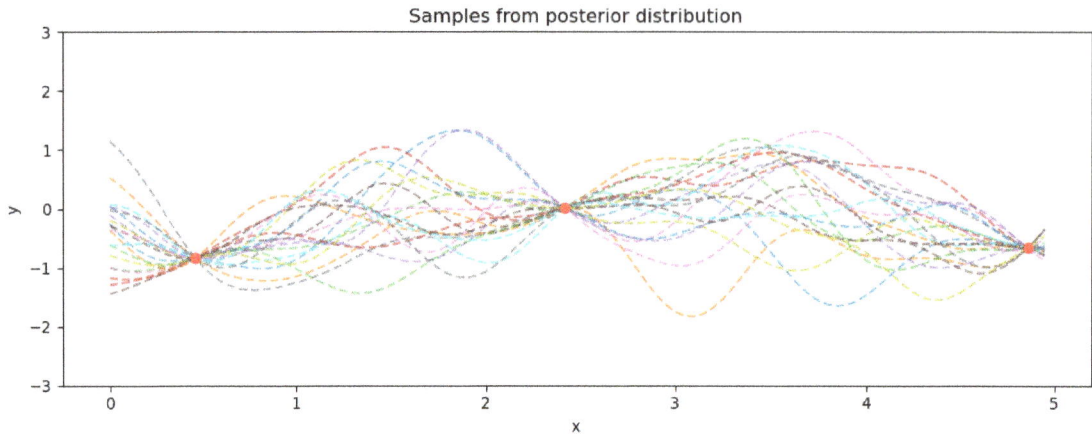

Figure 2.11: Illustration of space of possible functions after seeing some data

Here, we see that our possible functions all pass through our observed data points, but outside of those data points, our functions take on a range of very different values. In a simple linear model, we don't care about these deviations in possible values: we're happy to interpolate from one data point to another, as we see in *Figure 2.12*:

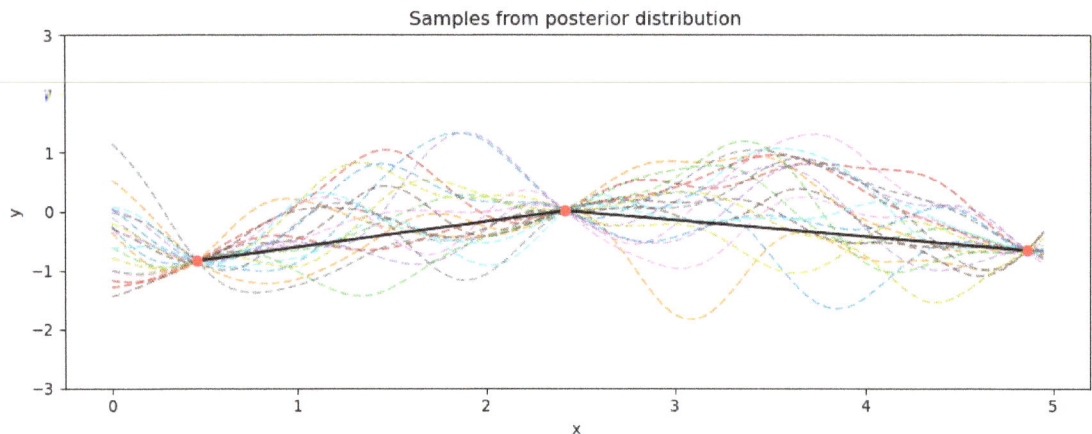

Figure 2.12: Illustration of linearly interpolating through our observations

But this interpolation can lead to wildly inaccurate predictions, and has no way of accounting for the degree of uncertainty associated with our model predictions. The deviations that we see here in the regions without data points are exactly what we want to capture with our GP. When there are a variety of possible values our function can take, then there is uncertainty – and through capturing the degree of uncertainty, we are able to estimate what the possible variation in these regions may be.

Formally, a GP can be defined as a function:

$$f(\mathbf{x}) \approx GP(m(\mathbf{x}), k(\mathbf{x}, \mathbf{x}')) \qquad (2.17)$$

Here, $m(\mathbf{x})$ is simply the mean of our possible function values for a given point $\mathbf{x}$:

$$m(\mathbf{x}) = \mathbb{E}[f(\mathbf{x})] \qquad (2.18)$$

The next term, $k(\mathbf{x}, \mathbf{x}')$ is a covariance function, or kernel. This is a fundamental component of the GP as it defines the way we model the relationship between different points in our data. GPs use the mean and covariance functions to model the space of possible functions, and thus to produce predictions as well as their associated uncertainties. Now that we've introduced some of the high-level concepts, let's dig a little deeper and understand exactly how it is they model the space of possible functions, and thus estimate uncertainty. To do this, we need to understand GP priors.

2.3.1   Defining our prior beliefs with kernels

GP kernels describe the prior beliefs we have about our data, and so you'll often see them referred to as GP priors. In the same way that the prior in equation 2.3 tells us something about the probability of the outcome of our two dice rolls, the GP prior tells us something important about the relationship we expect from our data.

While there are advanced methods for inferring a prior from our data, they are beyond the scope of this book. We will instead focus on more traditional uses of GPs, for which we select a prior using our knowledge of the data we're working with.

In the literature and any implementations you encounter, you'll see that the GP prior is often referred to as the **kernel** or **covariance function** (just as we have here). These three terms are all interchangeable, but for consistency with other work, we will henceforth refer to this as the kernel. Kernels simply provide a means of calculating a distance between two data points, and are expressed as $k(x, x')$, where x and x' are data points, and $k()$ represents the function of the kernel. While the kernel can take on many forms, there are a small number of fundamental kernels that are used in a large proportion of GP applications.

Perhaps the most commonly encountered kernel is the **squared exponential** or **radial basis function** (**RBF**) kernel. This kernel takes the form:

$$k(\mathbf{x}, \mathbf{x}') = \sigma^2 \exp -\frac{(\mathbf{x} - \mathbf{x}')^2}{2l^2} \tag{2.19}$$

This introduces us to a couple of common kernel parameters: l and σ^2. The output variance parameter σ^2 is simply a scaling factor, used to control the distance of the function from its mean. The length scale parameter l controls the smoothness of the function – in other words, how much your function is expected to vary across particular dimensions. This parameter can either be a scalar that is applied to all input dimensions, or a vector with a different scalar value for each input dimension. The latter is often achieved using **Automatic Relevance Determination**, or **ARD**, which identifies the relevant values in the input space.

GPs make predictions via a covariance matrix based on the kernel – essentially comparing a new data point to previously observed data points. However, just as with all ML models, GPs need to be trained, and this is where the length scale comes in. The length scale forms the parameters of our GP, and through the training process it learns the optimal value(s) for the length scale(s). This is typically done using a nonlinear optimizer, such as the **Broyden-Fletcher-Goldfarb-Shanno (BFGS)** optimizer. Many optimizers can be used, including optimizers you may be familiar with for deep learning, such as stochastic gradient descent and its variants.

Let's take a look at how different kernels affect GP predictions. We'll start with a straightforward example – a simple sine wave:

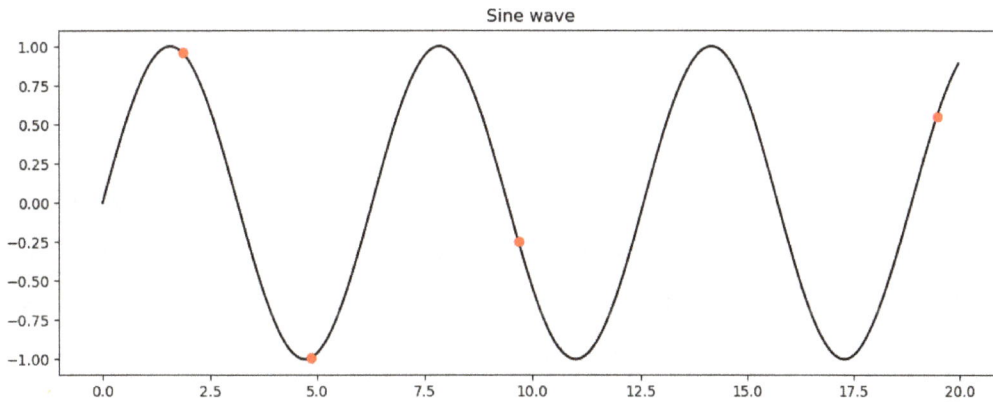

Figure 2.13: Plot of sine wave with four sampled points

We can see the function illustrated here, as well as some points sampled from this function. Now, let's fit a GP with a periodic kernel to the data. The periodic kernel is defined as:

$$k_{per}(\mathbf{x}, \mathbf{x}') = \sigma^2 \exp\left(\frac{2sin^2(\pi|\mathbf{x} - \mathbf{x}'|/p)}{l^2}\right) \tag{2.20}$$

Here, we see a new parameter: p. This is simply the period of the periodic function. Setting $p = 1$ and applying a GP with a periodic kernel to the preceding example, we get the following:

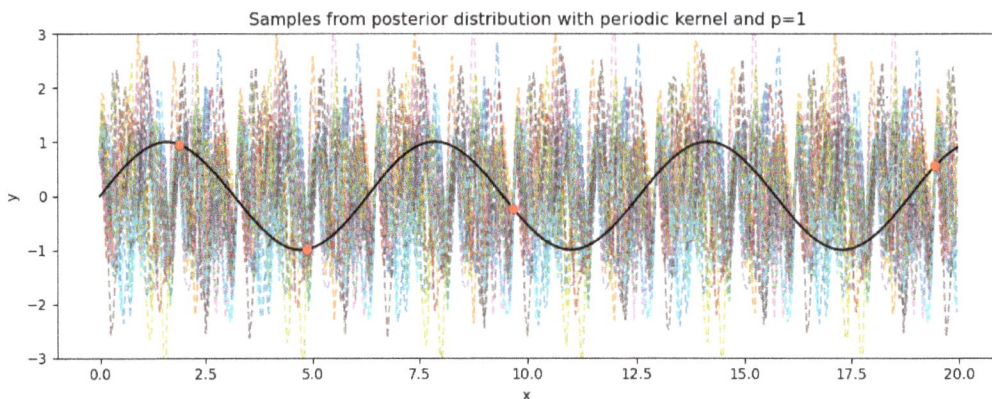

Figure 2.14: Plot of posterior predictions from a periodic kernel with $p = 1$

This looks pretty noisy, but you should be able to see that there is clear periodicity in the functions produced by the posterior. It's noisy for a couple of reasons: a lack of data, and a poor prior. If we're limited on data, we can try to fix the problem by improving our prior. In this case, we can use our knowledge of the periodicity of the function to improve our prior by setting $p = 6$:

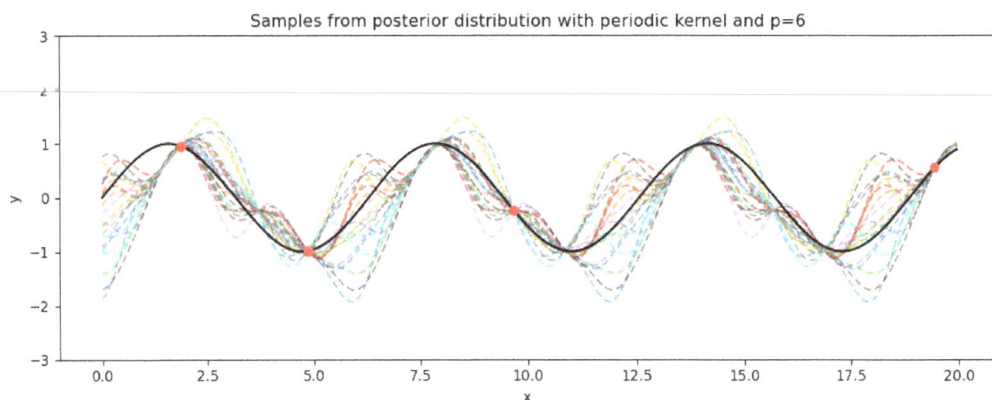

Figure 2.15: Plot of posterior predictions from a periodic kernel with $p = 6$

We see that this fits the data pretty well: we're still uncertain in regions for which we have little data, but the periodicity of our posterior now looks sensible. This is possible because we're using an informative prior; that is, a prior that incorporates information that describes the data well. This prior is composed of two key components:

- Our periodic kernel

- Our knowledge about the periodicity of the function

We can see how important this is if we modify our GP to use an RBF kernel:

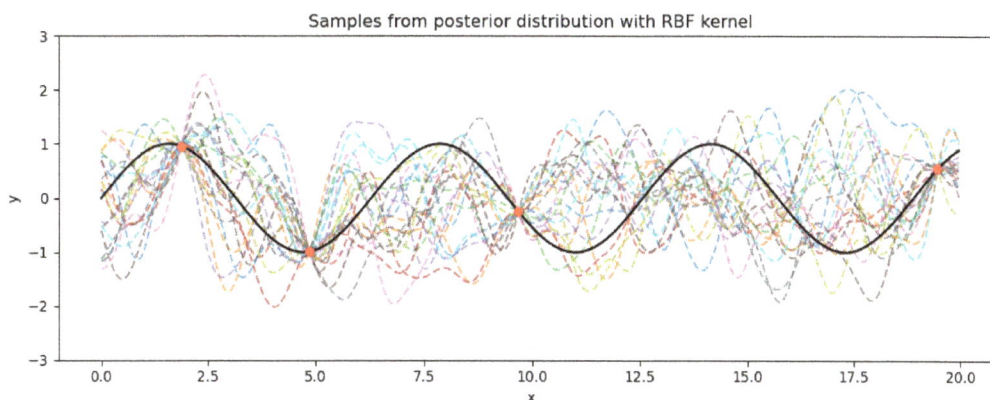

Figure 2.16: Plot of posterior predictions from an RBF kernel

With an RBF kernel, we see that things are looking pretty chaotic again: because we have limited data and a poor prior, we're unable to appropriately constrain the space of possible functions to fit our true function. In the ideal case, we'd fix this by using a more appropriate prior, as we saw in *Figure 2.15* – but this isn't always possible. Another solution is to sample more data. Sticking with our RBF kernel, we sample 10 data points from our function and re-train our GP:

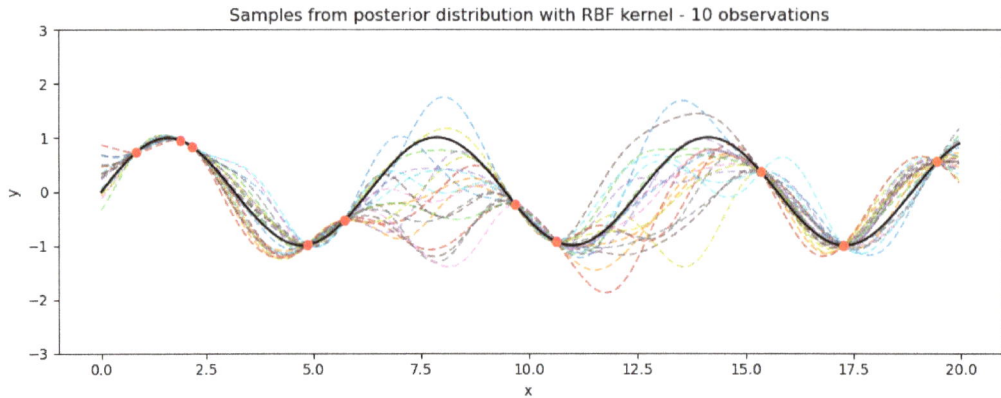

Figure 2.17: Plot of posterior predictions from an RBF kernel, trained on 10 observations

This is looking much better – but what if we have more data *and* an informative prior?

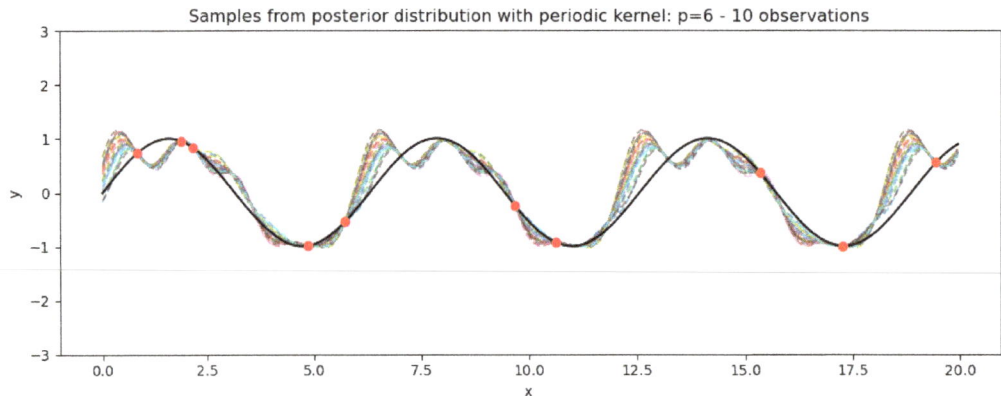

Figure 2.18: Plot of posterior predictions from a periodic kernel with $p = 6$, trained on 10 observations

The posterior now fits our true function very closely. Because we don't have infinite data, there are still some areas of uncertainty, but the uncertainty is relatively small.

Now that we've seen some of the core principles in action, let's return to our example from *Figures 2.10-2.12*. Here's a quick reminder of our target function, our posterior samples, and the linear interpolation we saw earlier:

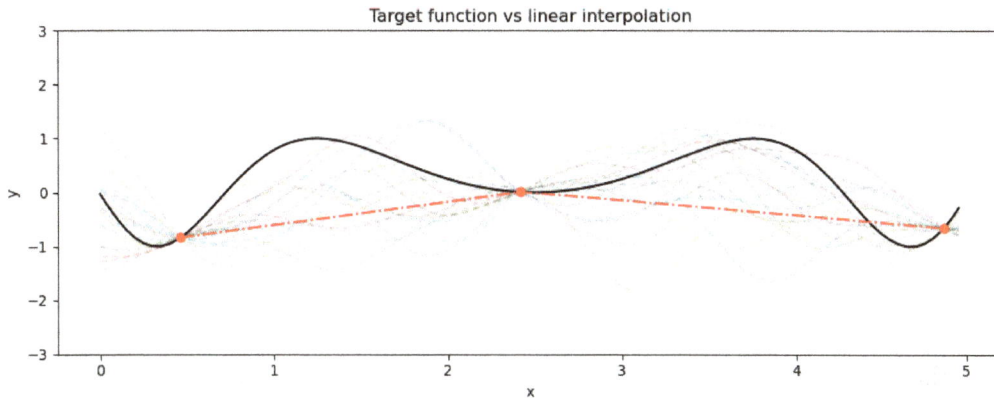

Figure 2.19: Plot illustrating the difference between linear interpolation and the true function

Now that we've got some idea of how a GP will affect our predictive posterior, it's easy to see that linear interpolation falls very short of what we achieve with a GP. To illustrate this more clearly, let's take a look at what the GP prediction would be for this function given the three samples:

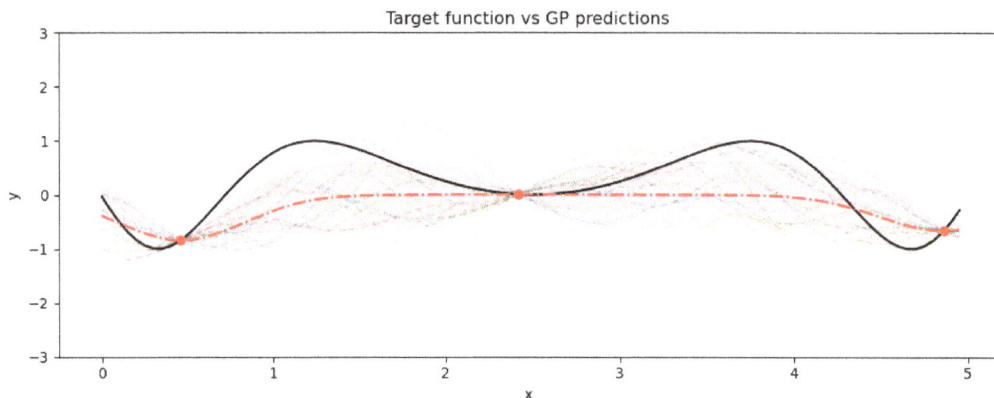

Figure 2.20: Plot illustrating the difference between GP predictions and the true function

Here, the dotted lines are our mean (μ) predictions from the GP, and the shaded area is the uncertainty associated with those predictions – the standard deviation (σ) around the mean. Let's contrast what we see in *Figure 2.20* with *Figure 2.19*. The differences may seem subtle at first, but we can clearly see that this is no longer a straightforward linear interpolation: the predicted values from the GP are being "pulled" toward our actual function values. As with our earlier sine wave examples, the behavior of the GP predictions are affected by two key factors: the prior (or kernel) and the data.

But there's another crucial detail illustrated in *Figure 2.20*: the predictive uncertainties from our GP. We see that, unlike many typical ML models, a GP gives us uncertainties associated with its predictions. This means we can make better decisions about what we do with the model's predictions – having this information will help us to ensure that our systems are more robust. For example, if the uncertainty is too great, we can fall back to a manual system. We can even keep track of data points with high predictive uncertainty so that we can continuously refine our models.

We can see how this refinement affects our predictions by adding a few more observations – just as we did in the earlier examples:

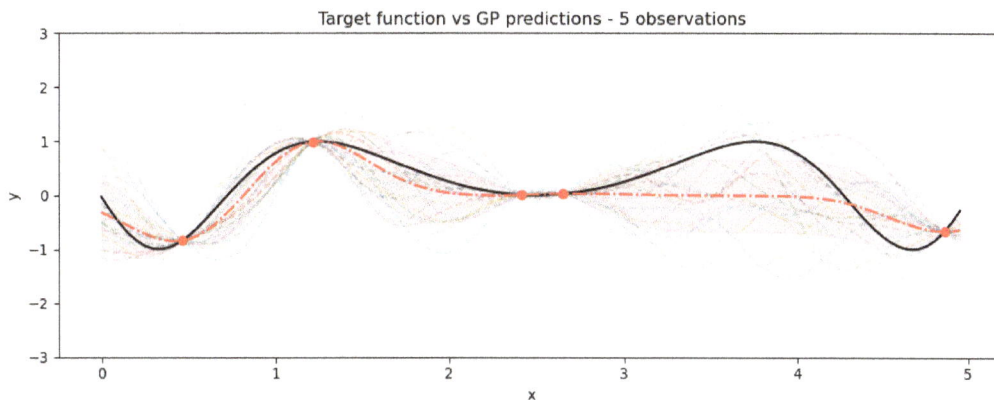

Figure 2.21: Plot illustrating the difference between GP predictions and the true function, trained on 5 observations

Figure 2.21 illustrates how our uncertainty changes over regions with different numbers of observations. We see here that between $x = 3$ and $x = 4$ our uncertainty is quite high. This makes a lot of sense, as we can also see that our GP's mean predictions deviate significantly from the true function values. Conversely, if we look at the region between $x = 0.5$ and $x = 2$, we can see that our GP's predictions follow the true function fairly closely, and our model is also more confident about these predictions, as we can see from the smaller interval of uncertainty in this region.

What we're seeing here is a great example of a very important concept: **well calibrated uncertainty** – also termed **high-quality uncertainty**. This refers to the fact that, in regions where our predictions are inaccurate, our uncertainty is also high. Our uncertainty estimates are **poorly calibrated** if we're very confident in regions with inaccurate predictions, or very uncertain in regions with accurate predictions.

GPs are what we can term a **well principled** method – this means that they have solid mathematical foundations, and thus come with strong theoretical guarantees. One of these guarantees is that they are well calibrated, and this is what makes GPs so popular: if we use GPs, we know we can rely on their uncertainty estimates.

Unfortunately, however, GPs are not without their shortcomings – we'll learn more about these in the following section.

2.3.2   Limitations of Gaussian processes

Given the fact that GPs are well-principled and capable of producing high-quality uncertainty estimates, you'd be forgiven for thinking they're the perfect uncertainty-aware ML model. GPs struggle in a few key situations:

- High-dimensional data

- Large amounts of data

- Highly complex data

The first two points here are largely down to the inability of GPs to scale well. To understand this, we just need to look at the training and inference procedures for GPs. While it's beyond the scope of this book to cover this in detail, the key point here is in the matrix operations required for GP training.

During training, it is necessary to invert a $D \times D$ matrix, where D is the dimensionality of our data. Because of this, GP training quickly becomes computationally prohibitive. This can be somewhat alleviated through the use of Cholesky deomposition, rather than direct matrix inversion. As well as being more computationally efficient, Cholesky decomposition is also more numerically stable. Unfortunately, Cholesky decomposition also has its weaknesses: computationally, its complexity is $O(n^3)$. This means that, as the size of our dataset increases, GP training becomes more and more expensive.

But it's not only training that's affected: because we need to compute the covariance between a new data point and all observed data points at inference, GPs have a $O(n^2)$ computational complexity at inference.

As well as the computational cost, GPs aren't light in memory: because we need to store our covariance matrix $\mathbf{K}$, GPs have a $O(n^2)$ memory complexity. Thus, in the case of large datasets, even if we have the compute resources necessary to train them, it may not be practical to use them in real-world applications due to their memory requirements.

The last point in our list concerns the complexity of data. As you are probably aware – and as we'll touch on in *Chapter 3, Fundamentals of Deep Learning* – one of the major advantages of DNNs is their ability to process complex, high-dimensional data through layers of non-linear transformations. While GPs are powerful, they're also relatively simple models, and they're not able to learn the kinds of powerful feature representations that are possible with DNNs.

All of these factors mean that, while GPs are an excellent choice for relatively low-dimensional data and reasonably small datasets, they aren't practical for many of the complex problems we face in ML. And so, we turn to BDL methods: methods that have the flexibility and scalability of deep learning, while also producing model uncertainty estimates.

2.4 Summary

In this chapter, we've covered some of the fundamental concepts and methods related to Bayesian inference. First, we reviewed Bayes' theorem and the fundamentals of probability theory – allowing us to understand the concept of uncertainty, as well as how we apply it to the predictions of ML models. Next, we introduced sampling, and an important class of algorithms: Markov Chain Monte Carlo, or MCMC, methods. Lastly, we covered Gaussian processes, and illustrated the crucial concept of well calibrated uncertainty. These key topics will provide you with the necessary foundation for the content that will follow, however, we encourage you to explore the recommended reading materials for a more comprehensive treatment of the topics introduced in this chapter.

In the next chapter, we will see how DNNs have changed the landscape of machine learning over the last decade, exploring the tremendous advantages offered by deep learning, and the motivation behind the development of BDL methods.

2.5   Further reading

There are a variety of techniques being explored to improve the flexibility and scalability of GPs – such as Deep GPs or Sparse GPs. The following resources explore some of these topics, and also provide a more thorough treatment of the content covered in this chapter:

- *Bayesian Analysis with Python*, Martin: this book comprehensively covers core topics in statistical modeling and probabilistic programming, and includes practical walk-throughs of various sampling methods, as well as a good overview of Gaussian processes and a variety of other techniques core to Bayesian analysis.

- *Gaussian Processes for Machine Learning*, Rasmussen and Williams: this is often considered the definitive text on Gaussian processes, and provides highly detailed explanations of the theory underlying Gaussian processes. A key text for anyone serious about Bayesian inference.

3

Fundamentals of Deep Learning

Throughout the book, when studying how to apply Bayesian methods and extensions to neural networks, we will encounter different neural network architectures and applications. This chapter will provide an introduction to common architecture types, thus laying the foundation for introducing Bayesian extensions to these architectures later on. We will also review some of the limitations of such common neural network architectures, in particular their tendency to produce overconfident outputs and their susceptibility to adversarial manipulation of inputs. By the end of this chapter, you should have a good understanding of deep neural network basics and know how to implement the most common neural network architecture types in code. This will help you follow the code examples found in later sections.

The content will be covered in the following sections:

- Introducing the multi-layer perceptron

- Reviewing neural network architectures

- Understanding the problem with typical neural networks

3.1 Technical requirements

To complete the practical tasks in this chapter, you will need a Python 3.8 environment with the pandas and scikit-learn stack and the following additional Python packages installed:

- TensorFlow 2.0

- Matplotlib plotting library

All of the code for this book can be found on the GitHub repository for the book: https://github.com/PacktPublishing/Enhancing-Deep-Learning-with-Bayesian-Inference.

3.2 Introducing the multi-layer perceptron

Deep neural networks are at the core of the deep learning revolution. The aim of this section is to introduce basic concepts and building blocks for deep neural networks. To get started, we will review the components of the **multi-layer perceptron (MLP)** and implement it using the TensorFlow framework. This will serve as the foundation for other code examples in the book. If you are already familiar with neural networks and know how to implement them in code, feel free to jump ahead to the *Understanding the problem with typical NNs* section, where we cover the limitations of deep neural networks. This chapter focuses on architectural building blocks and principles and does not cover learning rules and gradients. If you require additional background information for those topics, we recommend Sebastian Raschka's excellent *Python Machine Learning* book from Packt Publishing (in particular, *Chapter 2*).

The MLP is a feed-forward, fully connected neural network. Feed-forward means that the information in an MLP is only passed in one direction, from the input to the output layers; there are no backward connections. Fully connected means that each neuron is connected to all the neurons in the previous layer. To understand these concepts a bit better, let's have a look at *Figure* 3.1, which gives a diagrammatic overview of an MLP. In this example, the MLP has an **input layer** with three neurons (shown in red), two **hidden layers** with four neurons each (shown in blue), and one **output layer** with a single output node (shown in green). Imagine, for example, that we wanted to build a model that predicts housing prices in London. In this example, the three input neurons would represent values of three input features of our model, such as distance from the city centre, floor area, and the construction year of the house. As indicated by the black connections in the figure, these input values are then passed to and aggregated by each of the neurons of the first hidden layer. The values of these neurons are then, in turn, passed to and aggregated by the neurons in the second hidden layer and, finally, the output neuron, which will represent the house value predicted by our model.

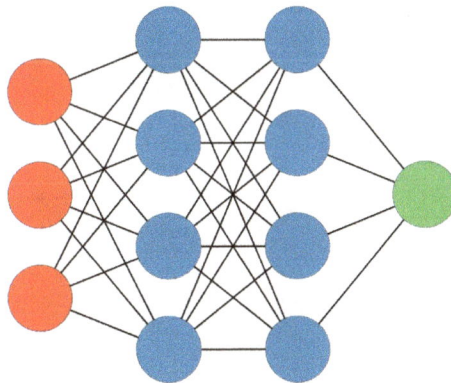

Figure 3.1: Diagram of a multi-layer perceptron

What does it mean exactly for a neuron to aggregate values? To understand this a bit better, let's focus on a single neuron and the operations that it performs on the values that are passed to it. In *Figure 3.2*, we have taken the network shown in *Figure 3.1* (left panel) and zoomed in on the first neuron in the first hidden layer and the neurons that pass values to it (central panel). In the right panel of the figure, we have slightly rearranged the neurons and have named the input neurons x_1, x_2, and x_3. We have also made the connections explicit, by naming the weights associated with them w_1, w_2, and w_3, respectively. From the right panel in the figure, we can see that an artificial neuron performs two essential operations:

1. First, it takes a weighted average over its inputs (indicated by the Σ).

2. Second, it takes the output of the first step and applies a non-linearity to it (indicated by the σ. Note that this does not indicate the standard deviation, which is what we'll use σ for throughout most of the book), such as a sigmoid function, for example.

The first operation can be expressed more formally as $z = \sum_{n=1}^{3} x_n w_n$. The second operation can be expressed as $a = \sigma(z) = \frac{1}{1+e^{-z}}$. The activation value of the neuron $a = \sigma(z)$ is then passed to the neurons in the second hidden layer, where the same operations are repeated.

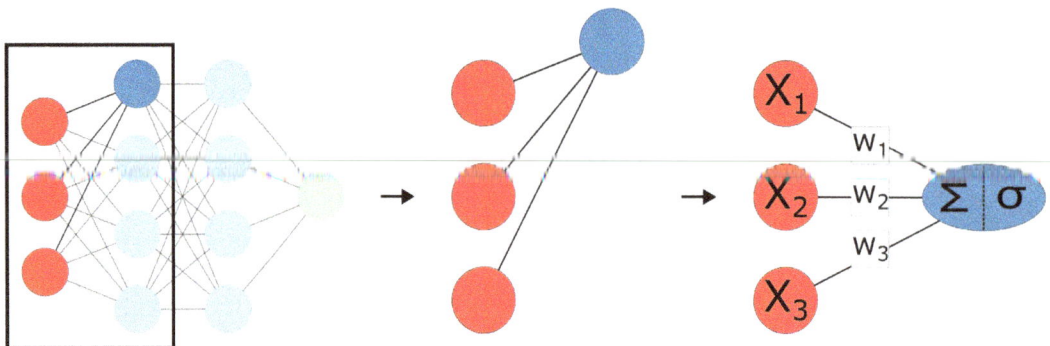

Figure 3.2: Aggregation and transformation performed by an artificial neuron in a neural network

Now that we have reviewed the different parts of an MLP model, let's implement one in TensorFlow. First, we need to import all the necessary functions. These include `Sequential` to build a feed-forward model such as MLP, `Input` to build the input layer, and `Dense` to build a fully-connected layer:

```
1  from tensorflow.keras.models import Sequential, Input, Dense
```

Equipped with these tools, implementing the MLP is a simple matter of chaining `Input` and `Dense` layers in the right order and with the right number of neurons:

```
1  multi_layer_perceptron = Sequential(
2      [
3          # input layer with 3 neurons
4          Input(shape=(3,))
5          # first hidden layer with 4 neurons
6          Dense(4, activation="sigmoid"),
7          # second hidden layer with 4 neurons
8          Dense(4, activation="sigmoid"),
9          # output layer
10         Dense(1, activation="sigmoid"),
11     ]
12 )
```

The aggregation in terms of weighted averaging is automatically handled by TensorFlow when using the `Dense` layer object. Furthermore, implementing an activation function becomes a simple matter of passing the name of the desired function to the `activation` parameter of the Dense layer (`sigmoid` in the preceding example).

Before we turn to other neural network architectures besides the MLP, a side note on the word *deep*. A neural network is considered deep if it has more than one hidden layer. The MLP shown previously, for example, has two hidden layers and can be considered a deep neural network. It is possible to add more and more hidden layers, creating very deep neural network architectures. Training such deep architectures comes with its own set of challenges and the science (or art) of training such deep architectures is called **deep learning (DL)**.

In the next section, we'll learn about some of the common deep neural network architectures and in the section thereafter, we will look at the practical challenges that come with them.

3.3   Reviewing neural network architectures

In the previous section, we saw how to implement a fully-connected network in the form of an MLP. While such networks were very popular in the early days of deep learning, over the years, machine learning researchers have developed more sophisticated architectures that work more successfully by including domain-specific knowledge (such as computer vision or **Natural Language Processing (NLP)**). In this section, we will review some of the most common of these neural network architectures, including **Convolutional Neural Networks (CNNs)** and **Recurrent Neural Networks (RNNs)**, as well as attention mechanisms and transformers.

3.3.1   Exploring CNNs

When looking back at the example of trying to predict London housing prices with an MLP model, the input features we used (distance to the city centre, floor area, and construction year of the house) were still "hand-engineered," meaning that a human looked at the problem and decided which inputs might be relevant to the model when making price predictions. What might such input features look like if we were trying to build a model that takes in images as input and tries to predict which object is shown in the image? One breakthrough moment for deep learning was the realization that neural networks can directly learn and extract the most useful features for a task from the raw data – in the case of visual object classification, these features are learned directly from the pixels in the image.

What would a neural network architecture need to look like if we wanted to extract the most relevant input features from an image for an object classification task? When trying to answer this question, early machine learning researchers turned to mammalian brains. Object classification is a task that our visual system performs relatively effortlessly. One observation that inspired the development of CNNs was that the visual cortex responsible for object recognition in mammals implements a hierarchy of feature extractors that work with increasingly large receptive fields. A **receptive field** is the area in the image that a biological neuron responds to. The neurons at the early layers of the visual cortex respond to relatively small regions of an image only, while neurons in layers higher up the hierarchy respond to areas that cover large parts (or even the entirety) of an input image.

Inspired by the cortical hierarchy in the brain, CNNs implement a hierarchy of feature extractors with artificial neurons higher up in the hierarchy having larger receptive fields. To understand how that works, let's look at how CNNs build features based on input images. *Figure 3.3* shows an early convolutional layer in a CNN operating on the input image (shown on the left) to extract features into a feature map (shown on the right). You can imagine the feature map as a matrix with n rows and m columns and every feature in the feature map as a scalar value. The example highlights two instances where the convolutional layer operates on different local regions of the image. In the first instance, the feature in the feature map receives input from the face of the kitten. In the second instance, the feature receives inputs from the kitten's right paw. The final feature map will be the result of repeating this same operation over all regions of the input image, sliding a kernel from left to right and from top to bottom to fill all the values in the feature map.

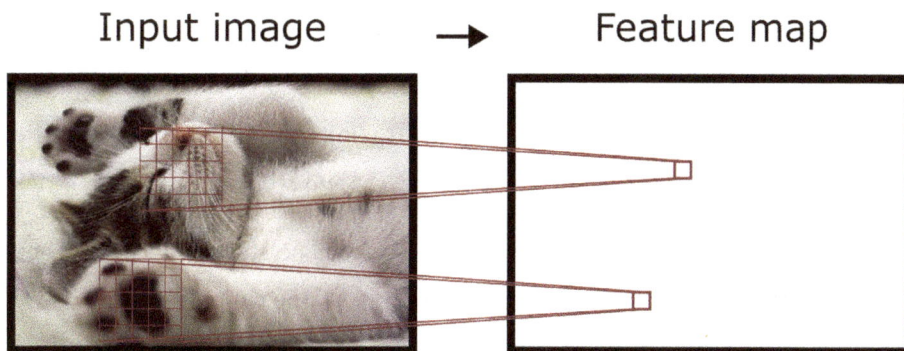

Figure 3.3: Building a feature map from the input image

What does such a single operation look like numerically? This is illustrated in *Figure 3.4.*

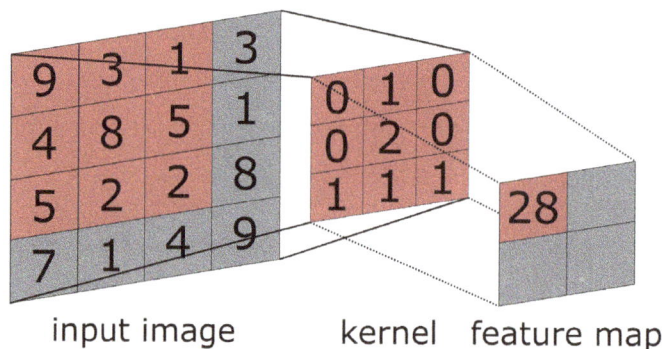

input image   kernel   feature map

Figure 3.4: Numerical operations performed by convolutional layer

Here, we have zoomed in on one part of the input image and made its pixel values explicit (left side). You can imagine that the kernel (shown in the middle) slides over the input image step by step. In the step that is shown in the figure, the kernel is operating on the upper-left corner of the input image (highlighted in red). Given the values in the input image and the kernel values, the final value in the feature map (**28** in the example) is obtained via weighted averaging: each value in the input image is weighted by the corresponding value in the kernel, which yields

$$9 * 0 + 3 * 1 + 1 * 0 + 4 * 0 + 8 * 2 + 5 * 0 + 5 * 1 + 2 * 1 + 2 * 1 = 28.$$

Slightly more formally, let us denote the input image by x and the kernel by w. Convolution in a CNN can then be expressed as $z = \sum_{i=1}^{n} \sum_{j=1}^{m} x_{i,j} w_{i,j}$. This is usually followed by a non-linearity, $a = \sigma(z)$, just like for the MLP. σ could be the sigmoid function introduced previously, but a more popular choice for CNNs is the **Rectified Linear Unit (ReLU)**, which is defined as $ReLU(z) = max(0, z)$.

In modern CNNs, many of these convolutional layers will be stacked on top of each other, such that the feature map that forms the output of one convolutional layer will serve as the input (image) for the next convolutional layer, and so forth. Putting convolutional layers in sequence like this allows the CNN to build more and more abstract feature representations. When studying feature maps at different positions of the hierarchy, it was shown by Matthew Zeiler et al. (see *Further reading*) that feature maps at early convolutional layers often show edges and simple textures, while feature maps at later convolutional layers show more complex patterns and parts of objects. Similar to the visual cortical hierarchy, neurons in later convolutional layers will tend to have larger receptive fields because they accumulate input from several earlier neurons, which in turn receive inputs from different local regions of the image.

The number of convolutional layers that are stacked on top of each other will determine the depth of a CNN: the more layers, the deeper the network. Another important dimension for a CNN is its width, which is determined by the number of convolutional kernels per layer. You can imagine that we can apply more than one kernel at a given convolutional layer, which will result in additional feature maps – one for every additional kernel. In this case, the kernels in the subsequent convolutional layer will need to be three-dimensional in order to handle the multitude of feature maps in the input, where the third dimension of the kernel will be determined by the number of incoming feature maps.

Along with convolutional layers, another common building block for CNNs is **pooling layers**, in particular **mean-pooling** and **max-pooling** layers. The function of these layers is to sub-sample the input, which reduces the input size of the image and thus the subsequent number of parameters needed in the network (and thus reduces the computational load and memory footprint).

How do pooling layers operate? In *Figure 3.5*, we see both a mean-pooling (left) and max-pooling (right) layer in operation. We see that, like convolutional layers, they operate on local regions of the input. The operations they perform are straightforward – either they take the mean or the maximum of the pixel values in their receptive field.

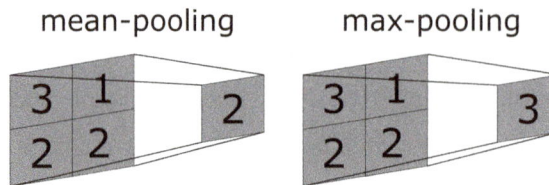

Figure 3.5: Numerical operations performed by pooling layers

In addition to computational and memory considerations, another advantage of pooling layers is that they can make the network more robust to small variations in the input. Imagine, for example, one of the input pixel values in the example changed to 0. This will either affect the output very little (mean-pooling layer) or not at all (max-pooling layer).

Now that we have reviewed the essential operations, let's implement a CNN in TensorFlow. Importing all the necessary functions includes the already familiar Sequential function to build a feed-forward model as well as the Dense layer. In addition, this time, we also import Conv2D for convolution and MaxPooling2D for max-pooling. With these tools, we can implement a CNN by chaining these layer functions in the right order:

```
1   from tensorflow.keras import Sequential
2   from tensorflow.keras.layers import Flatten, Conv2D, MaxPooling2D, Dense
3
4
5   convolutional_neural_network = Sequential([
6       Conv2D(32, (3,3), activation="relu", input_shape=(28, 28, 1)),
7       MaxPooling2D((2,2)),
8       Conv2D(64, (3,3), activation="relu"),
9       MaxPooling2D((2,2)),    Flatten(),
10      Dense(64, activation="relu"),
11      Dense(10)
12  ])
```

We have built a CNN by chaining a convolutional layer with 32 kernels, followed by a max-pooling operation, followed by a convolutional layer with 64 kernels, and another max-pooling operation. In the end, we add two Dense layers. The final Dense layer will serve to match the number of output neurons to the number of classes in a classification problem. In the preceding example, that number is 10. Our network is now ready for us to train.

CNNs have become crucial for a broad variety of problems, forming a key component in systems designed for a whole range of problems, from self-driving cars through to medical imaging. They also provided a foundation for other important neural network architectures, such as **graph convolutional networks (GCNs)**. But the field of deep learning wasn't able to dominate the world of machine learning with CNNs alone. In the next section, we'll learn about another important architecture: the recurrent neural network, an invaluable method for processing sequential data.

3.3.2 Exploring RNNs

The neural networks that we have seen so far are what we call feedforward networks: each layer of the network feeds into the next layer of the network; there is no cycle. Moreover, the convolutional neural networks we looked at receive a single input (an image) and output a single output: a label or a score for that label. But there are many cases in which we are working with something more complex than a single input, single output task. In this section, we will focus on a family of models called **recurrent neural networks (RNNs)**, which focus on processing sequences of inputs, with some also producing sequential outputs.

A typical example of an RNN task is machine translation. For example, translating the English sentence, "the apple is green," to French. For such a task to work, a network needs to consider the relationship between the inputs we feed it. Another task could be video classification, where we need to look at different frames of a video to classify the content of the video. An RNN processes an input one step at a time, where every time step can be denoted as t. At every time step, the model computes a hidden state h_t and an output y_t. But to compute h_t, the model does not only receive the input x_t but also the hidden state at the previous time step h_{t-1}. For a single time step, a vanilla RNN thus computes the following:

$$h_t = f(W_x x_t + W_h h_{t-1} + b) \tag{3.1}$$

Where:

- W_x are the weights of the RNN for the input x_t
- W_h are the weights for the hidden layer output from the previous time step h_{t-1}
- b is the bias term
- f is an activation function – in a vanilla RNN a *tanh* activation function

This way, at every time step, the model also has awareness of what happened at previous time steps because of the additional input h_{t-1}.

We can visualize the flow of an RNN as follows:

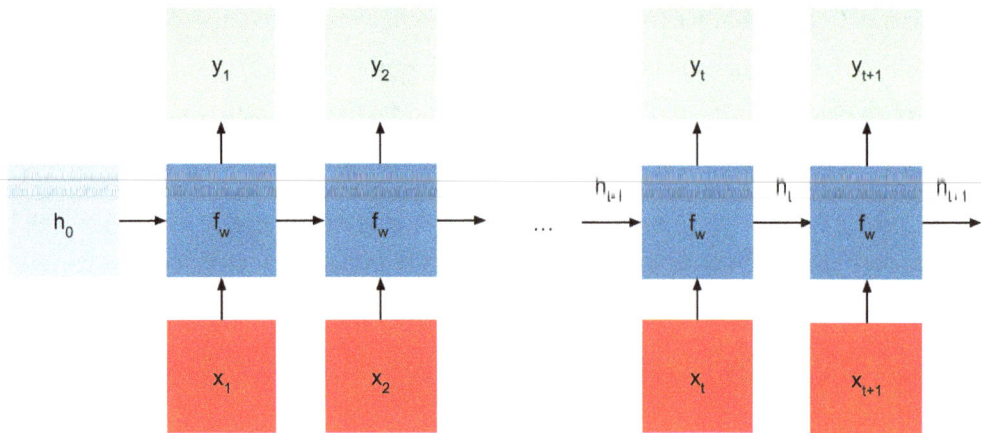

Figure 3.6: Example of an RNN

We can see that we need an initial hidden state as well at time step zero. This is usually just a vector of zeros.

One important variant of a vanilla neural network is a **sequence-to-sequence** **(seq2seq)** neural network, a popular paradigm in machine translation. The idea of this network is, as the name suggests, to take a sequence as input and output another sequence. Importantly, both sequences do not have to be of the same length. This enables the architecture to translate sentences in a more flexible way, which is crucial as different languages do not use the same number of words in sentences with the same meaning. This flexibility is achieved with an encoder-decoder architecture. This means that we will have two parts of our NN: an initial part that encodes the input up until a single weight (many inputs encoded to one hidden vector) matrix that is then used as input for the decoder of the network to produce multiple outputs (one input to many outputs). The encoder and the decoder have separate weight matrices. For a model with two inputs and two outputs, this can be visualized as follows:

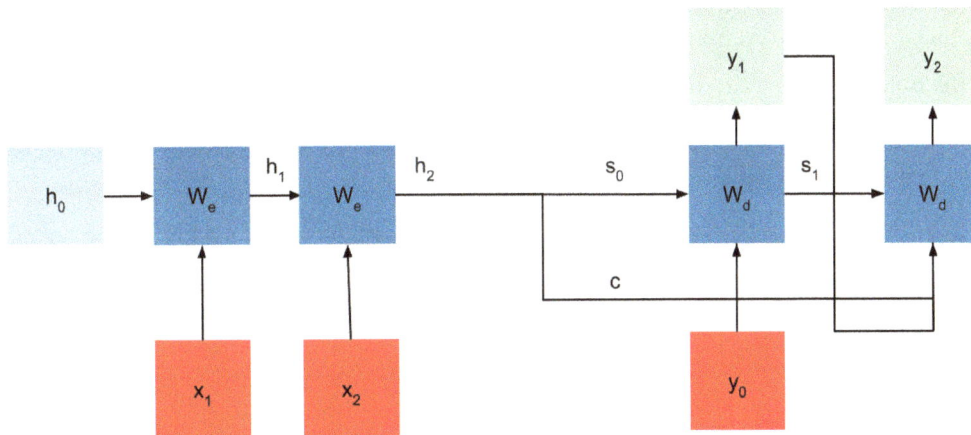

Figure 3.7: Example of a sequence-to-sequence network

In this figure w_e are the weights of the encoder and w_d are the weights of the decoder. We can see that compared to our RNN, we now have a new hidden state of the decoder s_0 and we can also observe c, which is a context vector. In standard sequence-to-sequence models, c is equal to the hidden state at the end of the encoder, whereas s_0, the initial hidden state of the encoder, is typically computed with one or more feedforward layers. The context vector is an additional input to each part of the decoder; it allows each part to use the information of the encoder.

3.3.3 Attention mechanisms

Although recurrent neural network models can be powerful, they have an important disadvantage: all information that the encoder can give to the decoder has to be in the hidden bottleneck layer – the hidden input state that the decoder receives at the start. That is fine for short sentences, but you can imagine that this becomes more difficult when we want to translate an entire paragraph or a very long sentence. We simply cannot expect that a single vector contains all information that is required to translate a long sentence. This downside is solved by a mechanism called attention. Later on, we will generalize the concept of attention, but let us first see how attention can be applied in the context of a seq2seq model.

Attention allows the decoder of a seq2seq model to "attend" to hidden states of the encoder according to some attention weights. This means that instead of having to rely on the bottleneck layer to translate the input, the decoder can go back to each hidden state of the encoder and decide how much information it wants to use. This is done via a context vector at every time step of the decoder that now functions as a probability vector, determining how much weight to give to each of the hidden states of the encoder. We can think of attention in this context as the following sequence for every time step of the decoder:

- $e_{t,i} = f(s_{t-1}, h_i)$ computes *alignment scores* for every hidden state of the encoder. This computation can be an MLP for every hidden state of the encoder, taking as input the current hidden state of the decoder s_{t-1}, and h_i the hidden states of the encoder.

- $e_{t,i}$ gives us alignment scores; they tell us something about the relation of every hidden state in the encoder and a single hidden state of the decoder. But the output of f is a scalar, which makes it impossible to compare different alignment scores. That is why we then take the softmax over all alignment scores to get a probability vector; attention weights: $a_{t,i} = softmax(e_{t,i})$. These weights are now values between 0 and 1 and tell us, for a single hidden state in the decoder, how much weight to give to every hidden state of the decoder.

- With our attention weights, we now take a weighted average of the hidden states of the encoder. This produces the context vector c_1, which can be used for the first time step of the decoder.

Because this mechanism computes attention weights for every time step of the decoder, the model is now much more flexible: at every time step, it knows how much weight to give to each part of the encoder. Moreover, because we are using an MLP here to compute the attention weights, this mechanism can be trained end to end.

This tells you a way to use attention in a sequence-to-sequence model. But the attention mechanism can be generalized to make it even more powerful. This generalization is used as a building block in the most powerful neural network applications you see today. It uses three main components:

- Queries, denoted as Q. You can think of these as the hidden states of the decoder.
- Keys, denoted as K. You can think of the keys as the hidden state of the inputs h_i.
- Values, denoted as V. In the standard attention mechanism, these are the same as the keys, but just separated as a separate value.

Together, the queries, keys, and values form the attention mechanism in the form of

$$Attention(Q, K, V) = softmax(\frac{QK^T}{\sqrt{d_k}})V \qquad (3.2)$$

We can distinguish three generalizations:

- Using an MLP to compute the attention weights is a relatively heavy operation for every time step. Instead, we can use something more lightweight that allows us to compute the attention weights for every hidden state of the decoder much faster: we use a scaled dot product of the hidden state of the decoder and the hidden states of the encoder. We scale the dot product by the square root of the dimension of K:

$$\frac{QK^T}{\sqrt{d_k}} \qquad (3.3)$$

This is because of two reasons:

- The softmax operation can lead to extreme values – values very close to zero and very close to one. This makes the optimization process more difficult. By scaling the dot product, we avoid this issue.

- The attention mechanism takes the dot product of vectors with a high dimension. This causes the dot product to be very large. By scaling the dot product, we counteract this tendency.

- We use the input vectors separately – we separate them as keys and values in different input streams. This gives the model more flexibility to handle them in different ways. Both are learnable matrices, so the model can optimize both in different ways.

- Attention takes a set of inputs as the query vector. This is more computationally efficient; instead of computing the dot product for every single query vector, we can do this for all of them at once.

These three generalizations make attention a very widely applicable algorithm. You see it in most of today's most performant models, some of the best image classification models – large language models that generate very realistic text or text-to-image models that can create the most beautiful and creative images. Because of the wide use of the attention mechanism, it is easily available in TensorFlow and other deep learning libraries. In TensorFlow, you can use attention like so:

```
1  from tensorflow.keras.layers import Attention
2  attention = Attention(use_scale=True, score_mode='dot')
```

And it can be called with our query, key, and value:

```
1  context_vector, attention_weights = attention(
2      inputs = [query, value, keys],
3      return_attention_scores = True,
4  )
```

In the preceding sections, we discussed some important building blocks of neural networks; we discussed the basic MLP, the concept of convolution, recurrent neural networks, and attention. There are other components we did not discuss here, and even more variants and combinations of the components we discussed. If you want to know more about these building blocks, refer to the reading list at the end of this chapter. There are excellent resources available in case you want to dive deeper into neural network architectures and components.

3.4   Understanding the problem with typical neural networks

The deep neural networks we discussed in previous sections are extremely powerful and, paired with appropriate training data, have enabled big strides in machine perception. In machine vision, convolutional neural networks enable us to classify images, locate objects in images, segment images into different segments or instances, and even to generate entirely novel images. In natural language processing, recurrent neural networks and transformers have allowed us to classify text, to recognize speech, to generate novel text or, as reviewed previously, to translate between two different languages.

However, these standard types of neural network models also have several limitations. In this section, we will explore some of these limitations. We will look at the following:

- How the prediction scores of such neural network models can be overconfident

- How such models can produce very confident predictions on OOD data

- How tiny, imperceptible changes in an input image can cause a model to make completely wrong predictions

3.4.1   Uncalibrated and overconfident predictions

One problem with modern vanilla neural networks is that they often produce outputs that are not well-calibrated. This means that the confidence scores produced by these networks no longer represent their empirical correctness. To understand better what that means, let's look at the reliability plot for the ideally-calibrated network that is shown in *Figure 3.8.*

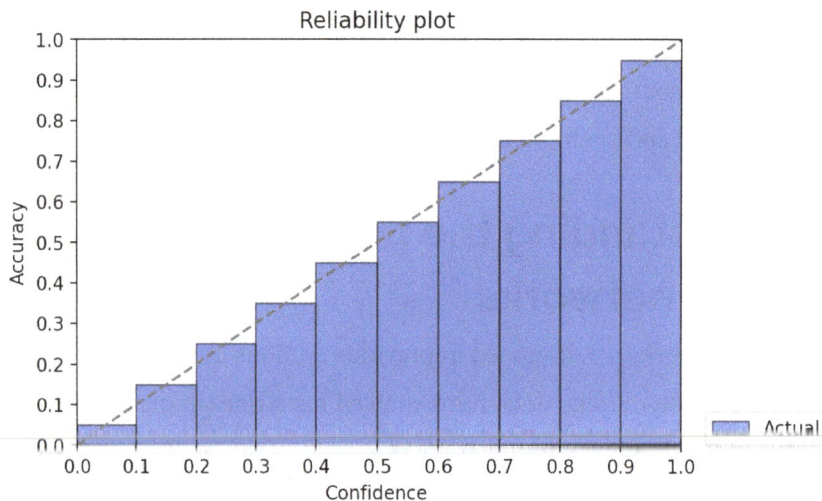

Figure 3.8: Reliability plot for a well-calibrated neural network. The empirically determined ("actual") accuracy is consistent with the prediction values output by the network

As you can see, the reliability plot shows accuracy (on the y axis) as a function of confidence (on the x axis). The basic idea is that, for a well-calibrated network, the output (or confidence) score associated with a prediction should match its empirical correctness. Here, empirical correctness is defined as the accuracy of a group of samples that all share similar output values and, for that reason, are grouped into the same bin in the reliability plot. So, for example, for the group of samples that all were assigned an output score between 0.7 and 0.8, the expectation, for a well-calibrated network, would be that 75% of these predictions should be correct.

To make this idea a bit more formal, let's imagine that we have a dataset $\mathbf{X}$ with $\mathbf{N}$ data samples. Each data sample $\mathbf{x}$ has a corresponding target label $\mathbf{y}$. In a classification setting, we would have y, which represents the membership to one in K classes. To obtain a reliability plot, we would use a neural network model to run inference over the entire dataset $\mathbf{X}$ to obtain an output score $\hat{y}$ per sample $\mathbf{x}$. We would then use the output scores to assign each data sample to a bin m in the reliability plot. In the preceding figure, we have opted for $M = 10$ bins. B_m would be the set of indices of samples that fall into bin m. Finally, we would measure and plot the average accuracy of the predictions for all the samples in a given bin, which is defined as

$acc(B_m) = \frac{1}{|B_m|} \sum_{i \in B_m} 1(\hat{y}_i = y_i).$

In the case of a well-calibrated network, the average accuracy of the samples in a given bin should match the confidence values of that bin. In the preceding figure, we can see, for example, that for samples that fell in the bin for output scores between 0.2 and 0.3, we observed a matching average accuracy of 0.25. Now let's see what happens in the case of an uncalibrated network that is over-confident in its predictions. Figure 3.9 illustrates this scenario, which is representative of the behavior shown by many modern vanilla neural network architectures.

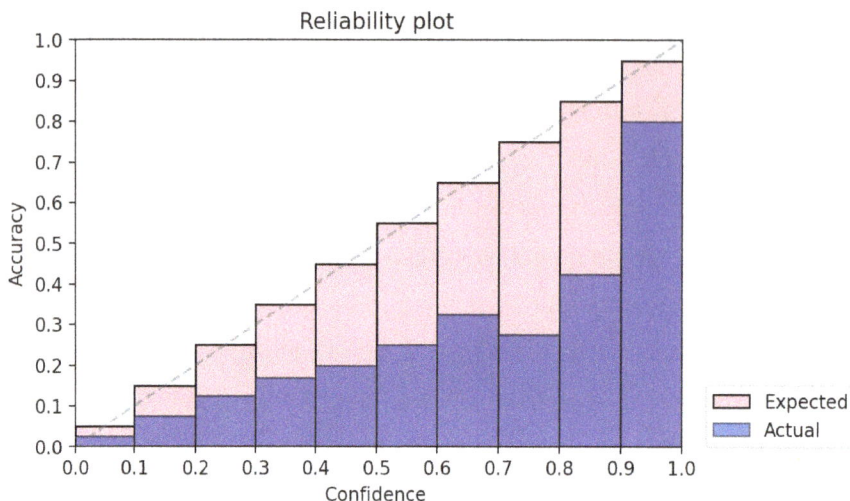

Figure 3.9: Reliability plot for an overconfident neural network. The "actual" empirically determined accuracy (purple bars) is consistently below the accuracy suggested by the prediction values of the network (pink bars and gray dashed line)

We can observe that for all the bins, the observed ("actual") accuracy for the samples in the bin is below the accuracy that is expected based on the output score of the samples. This is the expression of over-confident predictions. The network's output makes us believe that it has a high degree of confidence in its predictions, while in reality the actual performance does not match up.

Over-confident predictions can be very problematic in safety- and mission-critical applications, such as medical decision making, self-driving cars, and legal or financial decisions. Networks with overconfident predictions will lack the ability to indicate to us humans (or other networks) when they are likely to be wrong. This lack of awareness can become dangerous, for example, when a network is used to help decide whether a defendant should be granted bail or not. Assume a network is presented with a defendant's data (such as past convictions, age, education level) and predicts with 95% confidence score that bail should not be granted. Presented with this output, a judge might falsely think that the model can be trusted and base their verdict largely on the model's output. By contrast, calibrated confidence outputs can indicate to what degree we can trust the model's output. If the model is uncertain, this indicates that there is something about the input data that isn't well represented in the model's training data – indicating that the model is more likely to make a mistake. Thus, well-calibrated uncertainties allow us to decide whether to incorporate the model's predictions in our decision making, or whether to ignore the model entirely.

Drawing and inspecting reliability plots is useful for visualizing calibration in a few neural networks. However, sometimes we want to compare the calibration performance across several neural networks, possibly each network using more than one configuration. In such cases where we need to compare many networks and settings, it is useful to summarize the calibration of a neural network in a scalar statistic. The **Expected Calibration Error (ECE)** represents such a summary statistic. For every bin in the reliability plot, it measures the difference between the observed accuracy, $acc(B_m)$, and the accuracy we would have expected based on the output score of the samples, $conf(B_m)$, which is defined as $\frac{1}{|B_m|} \sum_{i \in B_m} \hat{y}$. Then, it takes a weighted average across all bins, where the weight for each bin is determined by the number of samples in the bin:

$$ECE = \sum_{m=1}^{M} \frac{|B_m|}{n} |acc(B_m) - conf(B_m)| \qquad (3.4)$$

This provides a first introduction to ECE and how it is measured. We will revisit ECE in more detail in *Chapter 8* by providing a code implementation as part of the *Revealing dataset shift with Bayesian methods* case study.

In the case of a perfectly calibrated neural network output, the ECE would be zero. The more uncalibrated a neural network is, the larger ECE would become. Let us look at some of the reasons for which neural networks are poorly calibrated and overconfident.

One reason is that the softmax function, which is usually the last operation of a classification network, uses the exponential function to make sure that all values are positive:

$$\sigma(\vec{z})_i = \frac{e^{z_i}}{\sum_{j=1}^{K} e^{z_j}} \tag{3.5}$$

The result of this is that small changes of the input to the softmax function can lead to substantial changes in its output.

Another reason for overconfidence is the increased model capacity of modern deep neural networks (Chuan Guo, et al. (2017)). Neural network architectures have become deeper (more layers) and wider (more neurons per layer) over the years. Such deep and wide neural networks have high variance and can very flexibly fit large amounts of input data. When experimenting with either the number of layers for a neural network or the number of filters per layer, Chuan Guo et al. observed that mis-calibration (and thus overconfidence) becomes worse with deeper and wider architectures. They also found that using batch normalization or training with less weight decay had a negative impact on calibration. These observations point to the conclusion that increased model capacity for modern deep neural network contributes to their overconfidence.

Finally, overconfident estimates can result from choosing particular neural network components. It has been shown, for example, that fully connected networks that use ReLU functions lead to continuous piecewise affine classifiers ([?]). This, in turn, implies that it is always possible to find input samples for which the ReLU network will produce high-confidence outputs. This holds even for input samples that are unlike the training data, for which generalization performance might be poor and we would thus expect lower confidence outputs. Such arbitrarily high confidence predictions also apply to convolutional networks that use either max- or mean-pooling following the convolutional layers, or any other network that results in a piecewise affine classifier function.

The problem is inherent to such neural network architectures and can only be addressed by changing the architecture itself ([?]).

3.4.2 Predictions on out-of-distribution data

Now that we have seen that models can be overconfident and therefore uncalibrated, let's look at another problem of neural networks. Neural networks are typically trained under the assumption that our test and train data are drawn from the same distribution. In practice, however, that is not always the case. The data that a model sees when it is deployed in the real world can change. We call these changes dataset shifts and they are typically divided into three categories:

- **Covariate shift**: The feature distribution $p(x)$ changes while $p(y|x)$ is fixed

- **Open-set recognition**: New labels appear at test time

- **Label shift**: The distribution of labels $p(y)$ changes while $p(x|y)$ is fixed

Examples of the preceding items include the following:

- Covariate shift: A model is trained to recognize faces. The training data consists of faces of mostly young people. The model at test time sees faces of all ages.

- Open-set recognition: A model is trained to classify a limited number of dog breeds. At test time, the model sees more dog breeds than present in the training dataset.

- Label shift: A model is trained to predict different diseases, of which some are very rare at the time of training. However, over time, the frequency of a rare disease changes and it becomes one of the most frequently seen diseases at test time.

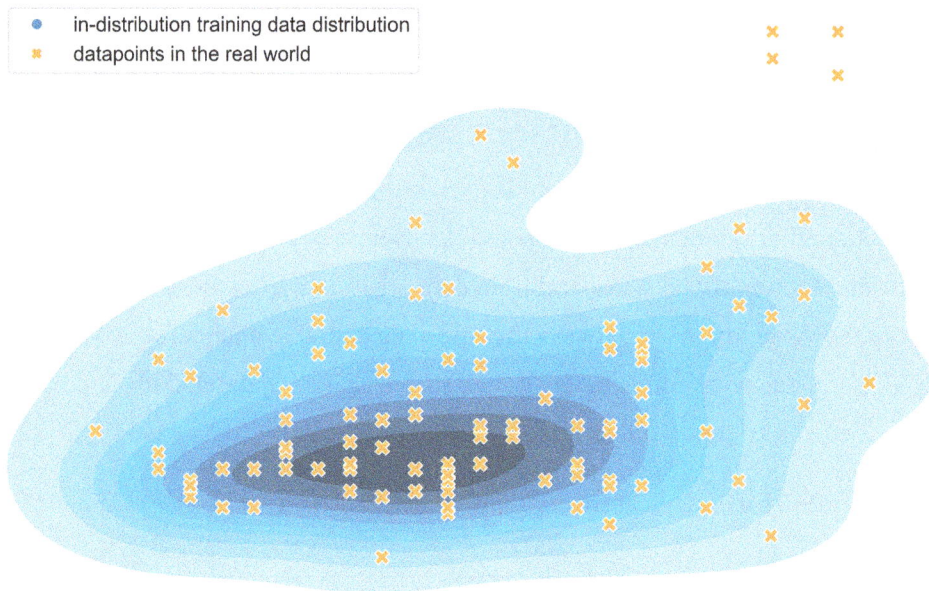

Figure 3.10: The training data distribution and data in the real world mostly overlap, but we cannot expect our model to perform well on the out-of-distribution points in the top right of the figure.

As a result of these changes, a model might be less performant when deployed in the real world if it is confronted with data that was not drawn from the same distribution as the training data. How likely a model is to be confronted with out-of-distribution data depends very much on the environment in which the model is deployed: some environments are more static (lower chance of out-of-distribution data), while others are more dynamic (higher chance of out-of-distribution data).

One reason for the problems of deep learning models with out-of-distribution data is that models often have a large number of parameters and can therefore memorize specific patterns and features of the training data instead of robust and meaningful representations of the data that reflect the underlying data distribution. When new data that looks slightly different from the training data presents itself at test time, the model does not actually have the ability to generalize and make the right prediction. One such example is an image of a cow (*Figure 3.11*) on a beach, when cows in the training dataset happened to be on green grasslands. Models often use the context present in the data to make predictions.

Figure 3.11: An object in a different environment (here, a cow on a beach) can make it difficult for a model to recognize that the image contains the object

Before we go over a practical example of how a simple model handles out-of-distribution data, let's examine a few approaches that can highlight the problem of out-of-distribution data in typical neural networks. Ideally, we would like our model to express high uncertainty when it encounters data that is different from the distribution it was trained on. If that is the case, out-of-distribution data will not be a big problem when a model is deployed in the real world. For example, in mission-critical systems where errors of a model are costly, there is often a certain confidence threshold that should be met before the prediction is trusted. If a model is well-calibrated and it assigns a low confidence score to out-of-distribution inputs, the business logic around the model can throw an exception and not use the model's output. For example, a self-driving car can alert the driver that it should take over control or it can slow down to avoid an accident.

However, common *neural networks do not know when they do not know*; they typically do not assign a low confidence score to out-of-distribution data.

An example of this is given in a Google paper titled *Can You Trust Your Model's Uncertainty? Evaluating Predictive Uncertainty Under Dataset Shift.* The paper shows that if you apply perturbations to a test dataset such as blur or noise, such that the images become more and more out-of-distribution, the accuracy of the model goes down.

However, the confidence calibration of the model also decreases. This means that the scores of the model are not trustworthy anymore on out-of-distribution data: they do not accurately indicate the model's confidence in its predictions. We will explore this behaviour ourselves in the *Revealing dataset shift with Bayesian methods* case study in *Chapter 8*.

Another way to determine how a model handles out-of-distribution data is by feeding it data that is not just perturbed, but completely different from the dataset it was trained on. The procedure to measure the model's out-of-distribution detection performance is then as follows:

1. Train a model on an in-distribution dataset.

2. Save the confidence scores of the model on the in-distribution test set.

3. Feed a completely different, out-of-distribution dataset to the model and save the corresponding confidence scores of the model.

4. Now, treat the scores from both datasets as scores from a binary problem: in-distribution or out-of-distribution. Compute binary metrics, such as the **area under the Receiver Operating Characteristic (AUROC)** curve or the area under the precision-recall curve.

This strategy tells you how well your model can separate in-distribution from out-of-distribution data. The assumption is that in-distribution data should always receive a higher confidence score than out-of-distribution data; in the ideal scenario, there is no overlap between the two distributions. In practice, however, this is not the case. Models do often give high confidence scores to out-of-distribution data. We will explore an example of this in the next section and a few solutions to this problem in later chapters.

3.4.3 Example of confident, out-of-distribution predictions

Let's see how a vanilla neural net can produce confident predictions on out-of-distribution data. In this example, we will first train a model and then feed it out-of-distribution data. To keep things simple, we will use a dataset with different types of dogs and cats and build a binary classifier that predicts whether the image contains a dog or a cat.

We first download our data:

```
1  curl -X GET https://s3.amazonaws.com/fast-ai-imageclas/oxford-iiit-pet.tgz  \
2  --output pets.tgz
3  tar -xzf pets.tgz
```

We then load our data into a dataframe:

```
1  import pandas as pd
2
3  df = pd.read_csv("oxford-iiit-pet/annotations/trainval.txt", sep=" ")
4  df.columns = ["path", "species", "breed", "ID"]
5  df["breed"] = df.breed.apply(lambda x: x - 1)
6  df["path"] = df["path"].apply(
7    lambda x: f"/content/oxford-iiit-pet/images/{x}.jpg"
8  )
```

We can then use scikit-learn's train_test_val() function to create a training and validation set:

```
1  import tensorflow as tf
2  from sklearn.model_selection import train_test_split
3
4  paths_train, paths_val, labels_train, labels_val = train_test_split(
```

```
5        df["path"], df["breed"], test_size=0.2, random_state=0
6    )
```

We then create our train and validation data. Our `preprocess()` function loads our
download image into memory and formats our label such that our model can process it.
We use a batch size of 256 and an image size of 160x160 pixels:

```
1   IMG_SIZE = (160, 160)
2   AUTOTUNE = tf.data.AUTOTUNE
3
4
5   @tf.function
6   def preprocess_image(filename):
7     raw = tf.io.read_file(filename)
8     image = tf.image.decode_png(raw, channels=3)
9     return tf.image.resize(image, IMG_SIZE)
10
11
12  @tf.function
13  def preprocess(filename, label):
14    return preprocess_image(filename), tf.one_hot(label, 2)
15
16
17  train_dataset = (tf.data.Dataset.from_tensor_slices(
18      (paths_train, labels_train)
19    ).map(lambda x, y: preprocess(x, y))
20    .batch(256)
21    .prefetch(buffer_size=AUTOTUNE)
22  )
23
24  validation_dataset = (tf.data.Dataset.from_tensor_slices(
25      (paths_val, labels_val))
26    .map(lambda x, y: preprocess(x, y))
```

```
27     .batch(256)
28     .prefetch(buffer_size=AUTOTUNE)
29  )
```

We can now create our model. To speed up learning, we can use transfer learning and start with a model that was pre-trained on ImageNet:

```
1  def get_model():
2    IMG_SHAPE = IMG_SIZE + (3,)
3    base_model = tf.keras.applications.ResNet50(
4        input_shape=IMG_SHAPE, include_top=False, weights='imagenet'
5    )
6    base_model.trainable = False
7    inputs = tf.keras.Input(shape=IMG_SHAPE)
8    x = tf.keras.applications.resnet50.preprocess_input(inputs)
9    x = base_model(x, training=False)
10   x = tf.keras.layers.GlobalAveragePooling2D()(x)
11   x = tf.keras.layers.Dropout(0.2)(x)
12   outputs = tf.keras.layers.Dense(2)(x)
13   return tf.keras.Model(inputs, outputs)
```

Before we can train the model, we first need to compile it. Compiling simply means that we specify a loss function and an optimizer for the model and, optionally, add some metrics for monitoring during training. In the following code, we specify that the model should be trained using the binary cross-entropy loss and the Adam optimizer and that, during training, we want to monitor the model's accuracy:

```
1  model = get_model()
2  model.compile(optimizer=tf.keras.optimizers.Adam(learning_rate=0.01),
3                loss=tf.keras.losses.BinaryCrossentropy(from_logits=True),
4                metrics=['accuracy'])
```

Because of transfer learning, just fitting our model for three epochs leads to a validation accuracy of about 99 percent:

```
1   model.fit(train_dataset, epochs=3, validation_data=validation_dataset)
```

Let's also test our model on the test set of this dataset. We first prepare our dataset:

```
1   df_test = pd.read_csv("oxford-iiit-pet/annotations/test.txt", sep=" ")
2   df_test.columns = ["path", "species", "breed", "ID"]
3   df_test["breed"] = df_test.breed.apply(lambda x: x - 1)
4   df_test["path"] = df_test["path"].apply(
5       lambda x: f"/content/oxford-iiit-pet/images/{x}.jpg"
6   )
7
8   test_dataset = tf.data.Dataset.from_tensor_slices(
9       (df_test["path"], df_test["breed"])
10  ).map(lambda x, y: preprocess(x, y)).batch(256)
```

We can then feed the dataset to our trained model. We obtain a test set accuracy of about 98.3 percent:

```
1   test_predictions = model.predict(test_dataset)
2   softmax_scores = tf.nn.softmax(test_predictions, axis=1)
3   df_test["predicted_label"] = tf.argmax(softmax_scores, axis=1)
4   df_test["prediction_correct"] = df_test.apply(
5       lambda x: x.predicted_label == x.breed, axis=1
6   )
7   accuracy = df_test.prediction_correct.value_counts(True)[True]
8   print(accuracy)
```

We now have a model that is pretty good at classifying cats and dogs. But what will happen if we give this model an image that is neither a cat or a dog? Ideally, the model should recognize that this image is not part of the data distribution and should output close to a uniform distribution. Let's see if this actually happens in practice. We feed our model some images from the ImageNet dataset – the actual dataset that was used to pre-train our model. The ImageNet dataset is large. That is why we download a subset of the dataset: a dataset called Imagenette. This dataset contains just 10 out of the 1,000 classes of the original ImageNet dataset:

```
1  curl -X GET https://s3.amazonaws.com/fast-ai-imageclas/imagenette-160.tgz \
2  --output imagenette.tgz
3  tar -xzf imagenette.tgz
```

We then take an image from the parachute class:

```
1  image_path = "imagenette-160/val/n03888257/ILSVRC2012_val_00018229.JPEG"
2  image = preprocess_image(image_path).numpy()
3  plt.figure(figsize=(5,5))
4  plt.imshow(image.astype(int))
5  plt.axis("off")
6  plt.show()
```

The image clearly does not contain a dog or a cat; it is obviously out-of-distribution:

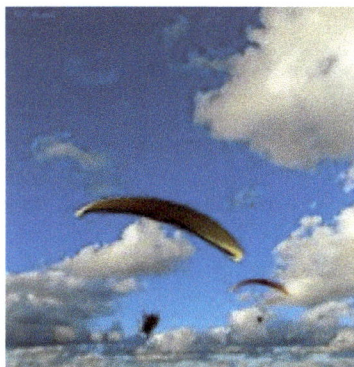

Figure 3.12: An image of a parachute from the ImageNet dataset

We run the image through our model and print the score:

```
1  logits = model.predict(tf.expand_dims(image, 0))
2  dog_score = tf.nn.softmax(logits, axis=1)[0][1].numpy()
3  print(f"Image classified as a dog with {dog_score:.4%} confidence")
4  # output: Image classified as a dog with 99.8226% confidence
```

We can see that the model classifies the image of a parachute as a dog with more than 99% confidence.

We can test the model's performance on the ImageNet parachute class more systematically as well. Let's run all the parachute images from the train split through the model and plot a histogram of the scores of the dog class.

We first create a small function to create special dataset with all the parachute images:

```
1  from pathlib import Path
2
3  parachute_image_dir = Path("imagenette-160/train/n03888257")
4  parachute_image_paths = [
5      str(filepath) for filepath in parachute_image_dir.iterdir()
6  ]
7  parachute_dataset = (tf.data.Dataset.from_tensor_slices(parachute_image_paths)
8  .map(lambda x: preprocess_image(x))
9  .batch(256)
10 .prefetch(buffer_size=AUTOTUNE))
```

We can then feed the dataset to our model and create a list of all the softmax scores related to the dog class:

```
1  predictions = model.predict(parachute_dataset)
2  dog_scores = tf.nn.softmax(predictions, axis=1)[:, 1]
```

We can then plot a histogram with these scores – this shows us the distribution of softmax scores:

```
1  plt.rcParams.update({'font.size': 22})
2  plt.figure(figsize=(10,5))
3  plt.hist(dog_scores, bins=10)
4  plt.xticks(tf.range(0, 1.1, 0.1))
5  plt.grid()
6  plt.show()
```

Ideally, we want these scores to be distributed close to 0.5 as the images in this dataset are neither dogs nor cats; the model should be very uncertain:

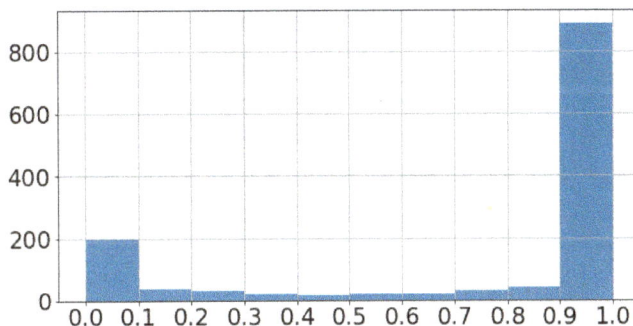

Figure 3.13: Distribution of softmax scores on the parachute dataset

However, we see something very different. More than 800 images are classified as a dog with at least 90% confidence. Our model clearly does not know how to handle out-of-distribution data.

3.4.4 Susceptibility to adversarial manipulations

One other vulnerability of most neural networks is that they are susceptible to adversarial attacks. Adversarial attacks, simply put, are ways to fool a deep learning system, most often in ways that would not fool humans. These attacks can be harmless or harmful. Here are some examples of adversarial attacks:

- A classifier can detect different types of animals. It classifies an image of a panda as a panda with 57.7% confidence. By slightly perturbing the image in a way that is invisible to humans, the image is now classified as a gibbon with 93.3% confidence.

- A model can detect whether a movie recommendation is positive or negative. It classifies a given movie as negative. By changing a word that does not change the overall tone of the review, for example, from "surprising" to "astonishing," the prediction of the model can change from a negative to a positive recommendation.

- A stop sign detector can detect stop signs. However, by putting a relatively small sticker on top of the stop sign, the model no longer recognises the stop sign.

These examples show that there are different kinds of adversarial attacks. A useful way to categorize adversarial attacks, is by trying to determine how much information about the model is available to the human (or machine) attacking the model. An attacker can always feed an input to the model, but what the model returns or how the attacker can inspect the model varies. With this lens, we can see the following categories:

- *Hard-label black box*: The attacker only has access to the labels resulting from feeding the model an input.

- *Soft-label black box*: The attacker has access to the scores and the labels of the model.

- *White box* setting: The attacker has full access to the model. They can access the weights and can see the scores, the structure of the model, and so on.

You can imagine that these different settings make it more or less difficult to attack a model. If people who want to fool a model can only see the label resulting from an input, they cannot be sure that a small change of the input will lead to a difference in the model's behavior as long as the label stays the same. This becomes easier when they have access to the model's label and scores. They can then see if a change in the input increases or decreases the confidence of the model. As a result, they can more systematically try to change our input, in ways that decrease the model's confidence in the label. This might make it possible to find a vulnerability of the model, given that there is enough time to change the input in an iterative fashion. Now, when someone has full access to the model (white box setting), finding a vulnerability might become even easier. They can now use more information to guide the change of the image, such as the gradient of the loss with respect to the input image.

The amount of information available to an attacker is not the only way to distinguish between different kinds of adversarial attacks; there are many types of attacks. For example, in the context of attacks on vision models, some attacks are based on single-patch adjustments of an image (or even single pixels!) while other attacks will change an entire image. Some attacks are specific to a single model, some attacks can be applied to multiple models. We can also distinguish between attacks that digitally manipulate an image and attacks that can be applied in the real world to fool a model, or attacks that are visible by the human eye and attacks that are not. As a result of the wide variety of attacks, the research literature about this topic is still very active – there are always new ways to attack a model, and subsequently a need to find defenses for these attacks.

In the section about out-of-distribution data, we trained a model to determine whether a given image is a cat or a dog. We saw that the classifier worked well: it achieved a test accuracy of about 98.3 percent. Is this model robust to adversarial attacks? Let's create an attack to find out. We will use the **fast-gradient sign method** (**FGSM**) to slightly perturb an image of a dog and make the model think it is actually an image of a cat. The fast-gradient sign method was introduced in 2014 by Ian Goodfellow et al. and remains one of the most famous examples of an adversarial attack. This is probably because of its simplicity; we will see that we will only need a few lines of code to create such an attack. Moreover, the results of this attack are astounding – Goodfellow himself mentioned that he could not really believe the results when he first tested this attack and he had to verify that the perturbed, adversarial image he fed to the model was actually different from the original input image.

To create an effective adversarial image, we have to make sure that the pixels in the image change – but only by so much that the change does not become apparent to the human eye. If we perturb the pixels in our image of a dog in such a way that the image now actually looks like a cat, then the model is not mistaken if it classifies that image as a cat. We make sure that we do not perturb the image too much by constraining the perturbation by the max norm – this essentially tells us that no pixel in the image can change by more than some amount ϵ:

$$\|\tilde{x} - x\|_\infty \leq \epsilon \tag{3.6}$$

Where $\tilde{x}$ is our perturbed image and x our original input image.

Now, to create our adversarial example in the fast-gradient sign method, we use the gradient of the loss with respect to our input image to create a new image. Instead of minimizing the loss as we want to do in gradient descent, we now want to maximize the loss. Given our network weights θ, input x, label y, and J as a function to compute our loss, we can create an adversarial image by perturbing the image in the following way:

$$\eta = \epsilon \operatorname{sgn} \left(\nabla_x J(\theta, x, y) \right) \tag{3.7}$$

In this equation, we compute the sign of the gradient of the loss with respect to the input, i.e. determine whether the gradient is positive (1), negative (-1) or 0. The sign enforces the max norm constraint, and by multiplying it by epsilon, we make sure that our perturbations are small – computing the sign just tells us that if we want to add or subtract epsilon in order to perturb the image in a way that hurts the model's performance on the image. η is now the perturbation that we want to add to our image:

$$\tilde{x} = x + \eta \tag{3.8}$$

Let's see what this looks like in Python. Given our trained network that classifies images as either dogs or cats, we can create a function that creates a perturbation that, when multiplied by epsilon and added to our image, creates an adversarial attack:

```python
import tensorflow as tf

loss_object = tf.keras.losses.BinaryCrossentropy()

def get_adversarial_perturbation(image, label):
    image = tf.expand_dims(image, 0)
    with tf.GradientTape() as tape:
        tape.watch(image)
        prediction = model(image)
        loss = loss_object(label, prediction)

    gradient = tape.gradient(loss, image)
    return tf.sign(gradient)[0]
```

We then create a small function that runs an input image through our model and returns the confidence of the model that the image contains a dog:

```python
def get_dog_score(image) -> float:
    scores = tf.nn.softmax(
        model.predict(np.expand_dims(image, 0)), axis=1
    ).numpy()[0]
    return scores[1]
```

We download an image of a cat:

```
curl https://images.pexels.com/photos/1317844/pexels-photo-1317844.jpeg > \
cat.png
```

Then, we pre-process it so we can feed it to our model. We set the label to 0, which corresponds to the cat label:

```
1  # preprocess function defined in the out-of-distribution section
2  image, label = preprocess("cat.png", 0)
```

We can perturb our image:

```
1  epsilon = 0.05
2  perturbation = get_adversarial_perturbation(image, label)
3  image_perturbed = image + epsilon * perturbation
```

Let's now get the confidence of the model that the original image is a cat, and the confidence of the model that the perturbed image is a dog:

```
1  cat_score_original_image = 1 - get_dog_score(image)
2  dog_score_perturbed_image = get_dog_score(image_perturbed)
```

With this in place, we can create the following plot, showing the original image, the perturbation applied to the image, and the perturbed image:

```
1   import matplotlib.pyplot as plt
2
3   ax = plt.subplots(1, 3, figsize=(20,10))[1]
4   [ax.set_axis_off() for ax in ax.ravel()]
5   ax[0].imshow(image.numpy().astype(int))
6   ax[0].title.set_text("Original image")
7   ax[0].text(
8       0.5,
9       -.1,
10      f"\"Cat\"\n {cat_score:.2%} confidence",
11      size=12,
```

```
12        ha="center",
13        transform=ax[0].transAxes
14    )
15    ax[1].imshow(perturbations)
16    ax[1].title.set_text(
17        "Perturbation added to the image\n(multiplied by epsilon)"
18    )
19    ax[2].imshow(image_perturbed.numpy().astype(int))
20    ax[2].title.set_text("Perturbed image")
21    ax[2].text(
22        0.5,
23        -.1,
24        f"\"Dog\"\n {dog_score:.2%} confidence",
25        size=12,
26        ha="center",
27        transform=ax[2].transAxes
28    )
29    plt.show()
```

Figure 3.14 shows both the original image and the perturbed image along with the model prediction for each image.

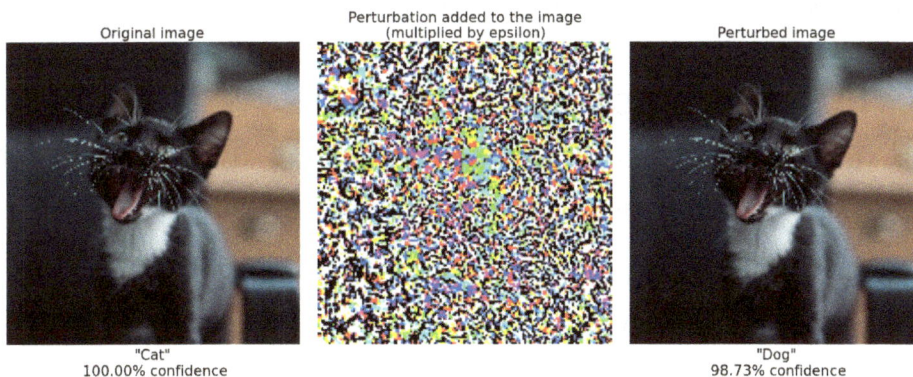

Figure 3.14: Example of an adversarial attack

In *Figure 3.14*, we can see that our model initially classifies the image as a cat, with a confidence of 100 percent. After we applied the perturbation (shown in the middle) to our initial cat image (shown on the left), our image (on the right) is now classified as a dog with 98.73 percent confidence, although the image visually looks the same as the original input image. We successfully created an adversarial attack that fools our model!

3.5   Summary

In this chapter, we have seen different types of common neural networks. First, we discussed the key building blocks of neural networks with a special focus on the multi-layer perceptron. Then we reviewed common neural network architectures: convolutional neural networks, recurrent neural networks, and the attention mechanism. All these components allow us to build very powerful deep learning models that can sometimes achieve super-human performance. However, in the second part of the chapter, we reviewed a few problems of neural networks. We discussed how they can be overconfident, and do not handle out-of-distribution data very well. We also saw how small, imperceptible changes to a neural network's input can cause the model to make an incorrect prediction.

In the next chapter, we will combine the concepts learned in this chapter and in *Chapter 3*, and discuss Bayesian deep learning, which has the potential to overcome some of the challenges of standard neural networks we have seen in this chapter.

3.6   Further reading

There are a lot of great resources to learn more about the essential building blocks of deep learning. Here are just a few popular resources that are a great start:

- Nielsen, M.A., 2015. *Neural networks and deep learning* (Vol. 25). San Francisco, CA, USA: Determination press., http://neuralnetworksanddeeplearning.com/.

- Chollet, F., 2021. *Deep learning with Python.* Simon and Schuster.

- Raschka, S., 2015. *Python Machine Learning.* Packt Publishing Ltd.

- Ng, Andrew, 2022, *Deep Learning Specialization.* Coursera.

- Johnson, Justin, 2019. EECS 498-007 / 598-005, *Deep Learning for Computer Vision.* University of Michigan.

To learn more about the problems of deep learning models, you can read some of the following resources:

- Overconfidence and calibration:

  - Guo, C., Pleiss, G., Sun, Y. and Weinberger, K.Q., 2017, July. *On calibration of modern neural networks.* In International conference on machine learning (pp. 1321-1330). PMLR.

  - Ovadia, Y., Fertig, E., Ren, J., Nado, Z., Sculley, D., Nowozin, S., Dillon, J., Lakshminarayanan, B. and Snoek, J., 2019. *Can you trust your model's uncertainty? evaluating predictive uncertainty under dataset shift.* Advances in neural information processing systems, 32.

- Out-of-distribution detection:

  - Hendrycks, D. and Gimpel, K., 2016. *A baseline for detecting misclassified and out-of-distribution examples in neural networks.* arXiv preprint arXiv:1610.02136.

  - Liang, S., Li, Y. and Srikant, R., 2017. *Enhancing the reliability of out-of-distribution image detection in neural networks.* arXiv preprint arXiv:1706.02690.

  - Lee, K., Lee, K., Lee, H. and Shin, J., 2018. *A simple unified framework for detecting out-of-distribution samples and adversarial attacks.* Advances in neural information processing systems, 31.

  - Fort, S., Ren, J. and Lakshminarayanan, B., 2021. *Exploring the limits of out-of-distribution detection.* Advances in Neural Information Processing Systems, 34, pp.7068-7081.

- Adversarial attacks:

    - Szegedy, C., Zaremba, W., Sutskever, I., Bruna, J., Erhan, D., Goodfellow, I. and Fergus, R., 2013. *Intriguing properties of neural networks.* arXiv preprint arXiv:1312.6199.

    - Goodfellow, I.J., Shlens, J. and Szegedy, C., 2014. *Explaining and harnessing adversarial examples.* arXiv preprint arXiv:1412.6572.

    - Nicholas Carlini, 2019. *Adversarial Machine Learning Reading List* `https://nicholas.carlini.com/writing/2019/all-adversarial-example-papers.html`.

You can have a look at the following resources to dive deeper into the topics and experiments covered in this chapter:

- Jasper Snoek, MIT 6.S191: *Uncertainty in Deep Learning*, January 2022.

- TensorFlow Core Tutorial, *Adversarial example using FGSM.*

- Goodfellow, I.J., Shlens, J. and Szegedy, C., 2014. *Explaining and harnessing adversarial examples.* arXiv preprint arXiv:1412.6572.

- Chuan Guo, Geoff Pleiss, Yu Sun, and Kilian Q Weinberger. On calibration of modern neural networks. In *International Conference on Machine Learning*, pages 1321–1330. PMLR, 2017.

- Stanford University School of Engineering, CS231N, *Lecture 16 | Adversarial Examples and Adversarial Training.*

- Danilenka, Anastasiya, Maria Ganzha, Marcin Paprzycki, and Jacek Mańdziuk, 2022. *Using adversarial images to improve outcomes of federated learning for non-IID data.* arXiv preprint arXiv:2206.08124.

- Szegedy, C., Zaremba, W., Sutskever, I., Bruna, J., Erhan, D., Goodfellow, I. and Fergus, R., 2013. *Intriguing properties of neural networks.* arXiv preprint arXiv:1312.6199.

- Sharma, A., Bian, Y., Munz, P. and Narayan, A., 2022. *Adversarial Patch Attacks and Defences in Vision-Based Tasks: A Survey.* arXiv preprint arXiv:2206.08304.

- Nicholas Carlini, 2019. *Adversarial Machine Learning Reading List* `https://nicholas.carlini.com/writing/2019/all-adversarial-example-papers.html`.

- Parkhi, O.M., Vedaldi, A., Zisserman, A. and Jawahar, C.V., 2012, June. *Cats and dogs.* In 2012 IEEE conference on computer vision and pattern recognition (pp. 3498-3505). IEEE. (Dataset cat vs dog).

- Deng, J., Dong, W., Socher, R., Li, L.J., Li, K. and Fei-Fei, L., 2009, June. *Imagenet: A large-scale hierarchical image database.* In 2009 IEEE conference on computer vision and pattern recognition (pp. 248-255). Ieee. (ImageNet dataset).

- Matthew D Zeiler and Rob Fergus. *Visualizing and understanding convolutional networks.* In European conference on computer vision, pages 818–833. Springer, 2014.

4

Introducing Bayesian Deep Learning

In *Chapter 2, Fundamentals of Bayesian Inference*, we saw how traditional methods for Bayesian inference can be used to produce model uncertainty estimates, and we introduced the properties of well-calibrated and well-principled methods for uncertainty estimation. While these traditional methods are powerful in many applications, *Chapter 2* also highlighted some of their limitations with respect to scaling. In *Chapter 3, Fundamentals of Deep Learning*, we saw the impressive things DNNs are capable of given large amounts of data; but we also learned that they aren't perfect. In particular, they often lack robustness for out-of-distribution data – a major concern when we consider the deployment of these methods in real-world applications.

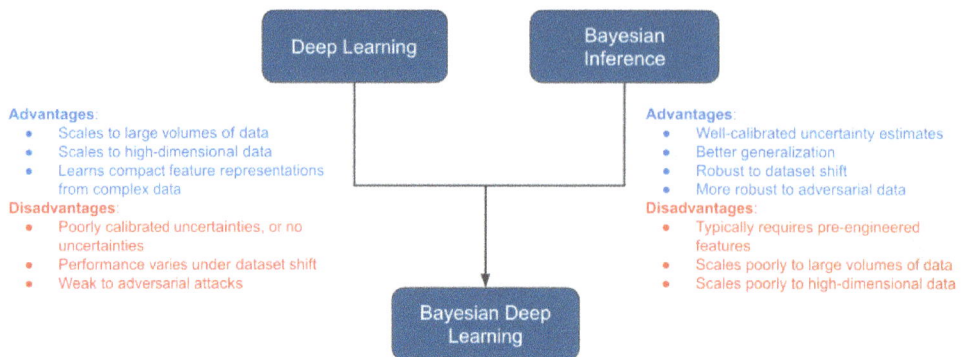

Figure 4.1: BDL combines the strengths of both deep learning and traditional Bayesian inference

BDL looks to ameliorate the shortcomings of both traditional Bayesian inference and standard DNNs, using the strengths from one method to address the weaknesses of the other. The fundamental idea is pretty straightforward: our DNNs gain uncertainty estimates, and so can be implemented more robustly, and our Bayesian inference methods gain the scalability and high-dimensional non-linear representation learning of DNNs.

While conceptually this is quite intuitive, practically it's not a case of just gluing things together. As the model complexity increases, so does the computational cost of Bayesian inference – making certain methods for Bayesian inference (such as via sampling) intractable.

In this chapter, we'll introduce the concept of an ideal **Bayesian Neural Network (BNN)** and discuss its limitations, and we'll learn about how we can use BNNs to create more robust deep learning systems. In particular, we'll be covering the following:

- The ideal BNN

- BDL fundamentals

- Tools for BDL

4.1 Technical requirements

To complete the practical tasks in this chapter, you will need a Python 3.8 environment with the `SciPy` stack and the following additional Python packages installed:

- TensorFlow 2.0

- TensorFlow Probability

- Seaborn plotting library

All of the code for this book can be found in the GitHub repository for the book: `https://github.com/PacktPublishing/Enhancing-Deep-Learning-with-Bayesian-Inference`.

4.2 The ideal BNN

As we saw in the previous chapter, a standard neural network comprises multiple layers. Each of these layers comprises a number of perceptrons – and these perceptrons comprise a multiplicative component (weight) and an additive component (bias). Each weight and bias parameter comprises a single parameter – or point estimate – and, in combination, these parameters transform the input to the perceptron. As we've seen, multiple layers of perceptrons are capable of achieving impressive feats when trained via backpropagation. However, these point estimates contain very limited information – let's take a look.

Generally speaking, the goal of deep learning is to find (potentially very, very many) parameter values that best map a set of inputs onto a set of outputs. That is, given some data, for each parameter in our network, we'll choose the parameter that best describes the data. This often boils down to taking the mean – or expectation – of the candidate parameter values. Let's see what this may look like for a single parameter in a neural network:

Input value	Ideal parameter	Predicted output value	Target output value	Error
4.80	5.00	(4.80 * 4.92) = 23.62	24.00	0.38
5.10	4.90	(5.10 * 4.92) = 25.09	24.99	0.10
4.90	4.80	(4.90 * 4.92) = 24.11	23.52	0.59
4.40	4.80	(4.40 * 4.92) = 21.65	21.12	0.51
5.20	5.10	(5.20 * 4.92) = 25.58	26.52	0.94
Mean	*4.92*	24.01	24.03	0.51
Standard deviation	0.12	1.37	1.78	0.27

Figure 4.2: A table of values illustrating how parameters are averaged in machine learning models

To understand this better, we'll use a table to illustrate the relationship between input values, model parameters, and output values. The table shows, for five example input values (first column), what the ideal parameter (second column) would be to obtain the target output value (fourth column). In this context, ideal here simply means that the input value multiplied by the ideal parameter will exactly equal the target output value. Because we need to find a single value that best maps our input data to our output data, we end up taking the expectation (or mean) of our ideal parameters.

As we see here, taking the mean of these parameters is the compromise our model needs to make in order to find a parameter value that best fits all five data points in the example. This is the compromise that is made with traditional deep learning – by using distributions, rather than point estimates, BDL can improve on this. If we look at our standard deviation (σ) values, we get an idea of how the variation in the *ideal* parameter values (and thus the variance in the input values) translates to a variation in the loss. So, what happens if we have a poor selection of parameter values?

Input value	Ideal parameter	Predicted output value	Target output value	Error
4.80	2.88	(4.80 * 6.38) = 30.60	24.00	6.60
5.10	6.40	(5.10 * 6.38) = 32.52	24.99	7.53
4.90	3.50	(4.90 * 6.38) = 31.24	23.52	7.72
4.40	3.20	(4.40 * 6.38) = 28.05	21.12	6.93
5.20	15.90	(5.20 * 6.38) = 33.16	26.52	6.64
Mean	6.38	31.11	24.03	7.08
Standard deviation	4.93	1.77	1.78	0.46

Figure 4.3: A table of values illustrating how parameter σ increases for poor sets of parameters

If we compare *Figure 4.2* and *Figure 4.3*, we see how a significant variance in parameter values can lead to poorer approximation from the model, and that larger σ can be indicative of an error (at least for well-calibrated models). While in practice things are a little more complicated, what we see here is essentially what's happening in every parameter of a deep learning model: parameter distributions are distilled down to point estimates, losing information in the process. In BDL, we're interested in harnessing the additional information from these parameter distributions, using it for more robust training and for the creation of uncertainty-aware models.

BNNs look to achieve this by modeling the distribution over neural network parameters. In the ideal case, the BNN would be able to learn any arbitrary distribution for every parameter in the network. At inference time, we would sample from the NN to obtain a distribution of output values. Using the sampling methods introduced in *Chapter 2*, we would repeat this process until we have obtained a statistically sufficient number of samples from which we could assume a good approximation of our output distribution. We could then use this output distribution to infer something about our input data, whether that be classifying speech content or performing regression on house prices.

Because we'd have parameter distributions, rather than point estimates, our ideal BNN would produce precise uncertainty estimates. These would tell us how likely the parameter values are given the input data. In doing so, they would allow us to detect cases where our input data deviates from the data seen at training time, and to quantify the degree of this deviation by how far a given sample of values lies from the distribution learned at training time. With this information, we would be able to handle our neural network outputs more intelligently – for example, if they're highly uncertain, then we could fall back to some safe, pre-defined behavior. This concept of interpreting model predictions based on uncertainties should be familiar: we saw this in *Chapter 2*, where we learned that high uncertainties are indicative of erroneous model predictions.

Looking back to *Chapter 2* again, we saw that sampling quickly becomes computationally intensive. Now imagine sampling from a distribution for each parameter in an NN – even if we take a relatively small network such as MobileNet (an architecture specifically designed to be more computationally efficient), we're still looking at an enormous 4.2 million parameters. Performing this kind of sampling-based inference on such a network would be incredibly computationally intensive, and this would be even worse for other network architectures (for example, AlexNet has 60 million parameters!).

Because of this intractability, BDL methods make use of various approximations in order to facilitate uncertainty quantification. In the next section, we'll learn about some of the fundamental principles applied to make uncertainty estimates possible with DNNs.

4.3 BDL fundamentals

Throughout the rest of the book, we will introduce a range of methods necessary to make BDL possible. There are a number of common themes present through these methods. We'll cover these here, so that we have a good understanding of these concepts when we encounter them later on.

These concepts include the following:

- **Gaussian assumptions**: With many BDL methods, we use Gaussian assumptions to make things computationally tractable

- **Uncertainty sources**: We'll take a look at the different sources of uncertainty, and how we can determine the contributions of these sources for some BDL methods

- **Likelihoods**: We were introduced to likelihoods in *Chapter 2*, and here we'll learn more about the importance of likelihood as a metric for evaluating the calibration of probabilistic models

Let's look at each of these in the following subsections.

4.3.1 Gaussian assumptions

In the ideal case described previously, we talked about learning distributions for each neural network parameter. While realistically each parameter would follow a specific non-Gaussian distribution, this would make an already difficult problem even *more* difficult. This is because, for a BNN, we're interested in learning two key probabilities:

- The probability of the weights W given some data D:

$$P(W|D) \tag{4.1}$$

- The probability of some output $\hat{y}$ given some input $\mathbf{x}$:

$$P(\hat{y}|\mathbf{x}) \tag{4.2}$$

Obtaining these probabilities for arbitrary probability distributions would involve solving intractable integrals. Gaussian integrals, on the other hand, have closed-form solutions – making them a very popular choice for approximating distributions.

For this reason, it's common in BDL to assume that we can closely approximate the true underlying distribution of our weights with Gaussian distributions (similarly to what we've seen in *Chapter 2*). Let's see what this would look like – taking our typical linear perceptron model:

$$z = f(x) = \beta X + \xi \tag{4.3}$$

Here, x is our input to the perceptron, β is our learned weight value, ξ is our learned bias value, and z is the value that is returned (typically passed to the next layer). With a Bayesian approach, we turn our parameters β and ξ into distributions, rather than point estimates, such that:

$$\beta \approx \mathcal{N}(\mu_\beta, \sigma_\beta) \tag{4.4}$$

$$\xi \approx \mathcal{N}(\mu_\xi, \sigma_\xi) \tag{4.5}$$

The learning process would now involve learning four parameters instead of two, as each Gaussian is described by two parameters: the mean (μ) and standard deviation (σ). Doing this for each perceptron in our neural network, we end up doubling the number of parameters we need to learn – we can see this illustrated, starting with *Figure 4.4*:

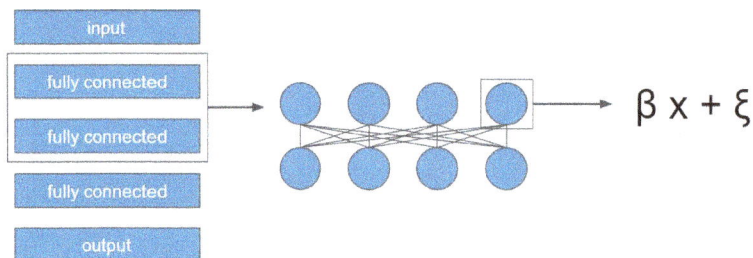

Figure 4.4: An illustration of a standard DNN

Introducing one-dimensional Gaussian distributions for our weights, our network becomes as follows:

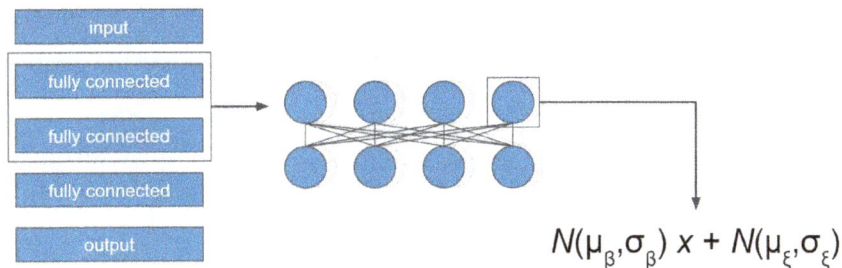

Figure 4.5: An illustration of a BNN with Gaussian priors over the weights

In *Chapter 5, Principled Approaches for Bayesian Deep Learning*, we'll see methods that do exactly this. While this does increase the computational complexity and memory footprint of our network, it makes the process of Bayesian inference with NNs manageable – making it a very worthwhile trade-off.

So, what is it we're actually trying to capture in these uncertainty estimates? In *Chapter 2*, we saw how uncertainty varies according to the sample of data used for training – but what are the sources of this uncertainty, and why is it important in deep learning applications? Let's continue on to the next section to find out.

4.3.2 Sources of uncertainty

As we saw in *Chapter 2*, and as we'll see later on in the book, we typically deal with uncertainties as scalar variables associated with a parameter or output. These variables represent the variation in the parameter or output of interest, but while they are just scalar variables, there are multiple sources contributing to their values. These sources of uncertainty fall into two categories:

- **Aleatoric uncertainty**, otherwise known as observational uncertainty or data uncertainty, is the uncertainty associated with our inputs. It describes the variation in our **observations**, and as such is **irreducible**.

- **Epistemic uncertainty**, otherwise known as model uncertainty, is the uncertainty that stems from our model. In the case of machine learning, this is the variance associated with the parameters of our model that *does not* stem from the observations, and is instead a product of the model, or how the model is trained. For example, in *Chapter 2*, we saw how different priors affected the uncertainty produced by Gaussian processes. This is an example of how model parameters influence the epistemic uncertainty – in this case, because they explicitly modify how the model interprets the relationship between different data points.

We can build an intuition of these concepts through some simple examples. Let's say we have a basket of fruit containing apples and bananas. If we measure the height and length of some apples and bananas, we'll see that apples are generally round, and that bananas are generally long, as illustrated in *Figure 4.6*. We know from our observations that the exact dimensions of each fruit varies: we accept that there is randomness, or stochasticity, associated with the measurements of any given distribution of apples, but we know that they will all be roughly similar. This is the **irreducible uncertainty**: the inherent uncertainty in the data.

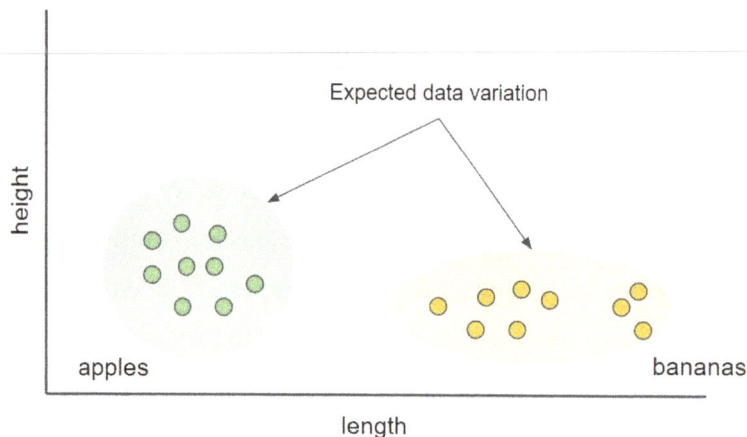

Figure 4.6: An illustration of aleatoric uncertainty, using fruit shapes as an example

We can make use of this information to build a model to classify fruit as either apples or bananas according to these input features. But what happens if we mainly train our model on apples, with only a few measurements for bananas? This is illustrated in *Figure 4.7.*

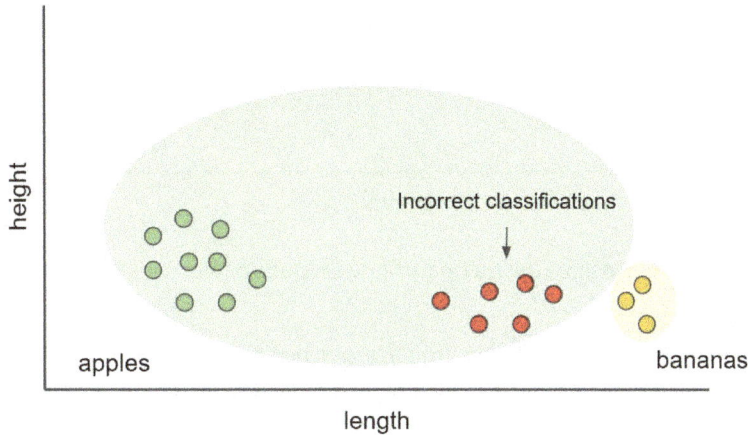

Figure 4.7: An illustration of high epistemic uncertainty based on our fruit example

Here, we see that – because of limited data – our model has incorrectly classified bananas as apples. While these data points fall within our model's apple boundary, we also see that they lie very far from the other apples, meaning that, although they're classified as apples, our model (if it's Bayesian) will have a high predictive uncertainty associated with these data points. This epistemic uncertainty is very useful in practical applications: it gives us an indication of when we can trust our model, and when we should be cautious about our model's predictions. Unlike aleatoric uncertainty, epistemic uncertainty is **reducible** – if we give our model more examples of bananas, its class boundaries will improve, and the epistemic uncertainty will approach the aleatoric uncertainty.

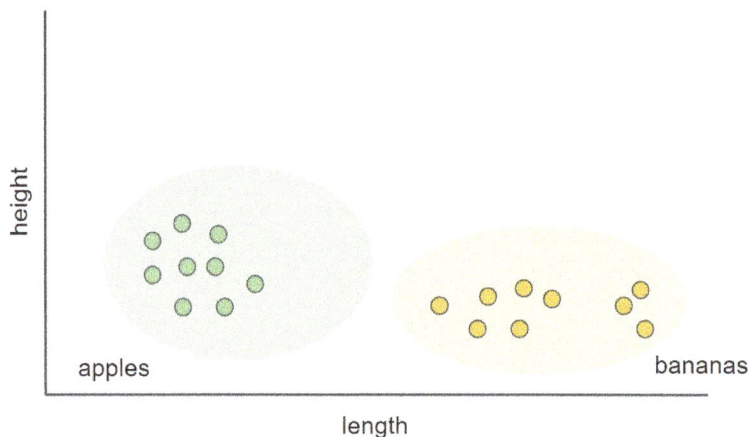

Figure 4.8: Illustration of low epistemic uncertainty

In *Figure 4.8*, we see that the epistemic uncertainty has reduced significantly now that our model has observed more data, and it's looking a lot more like the aleatoric uncertainty illustrated in *Figure 4.6*. Epistemic uncertainty is therefore incredibly useful, both for indicating how much we can trust our model, and as a means of improving our model's performance.

As deep learning approaches are increasingly applied in mission-critical and safety-critical applications, it's crucial that the methods we use can estimate the degree of epistemic uncertainty associated with their predictions. To illustrate this, let's change the domain of our example from *Figure 4.7*: instead of classifying fruit, we're now classifying whether a jet engine is operating within safe parameters, as shown in *Figure 4.9*.

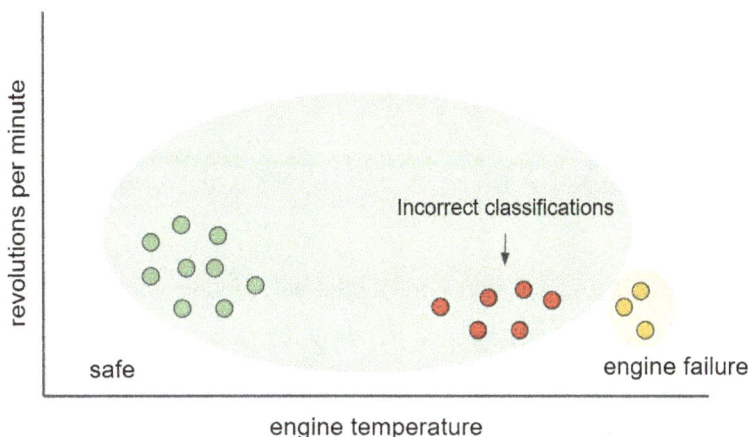

Figure 4.9: An illustration of high epistemic uncertainty in a safety-critical application

Here, we see that our epistemic uncertainty could be a life-saving indicator of engine failure. Without this uncertainty estimate, our model would assume that all is fine, even though the temperature of the engine is unusual given the other parameters – this could lead to catastrophic consequences. Fortunately, because of our uncertainty estimates, our model is able to tell us that something is wrong, despite the fact that it's never encountered this situation before.

Separating sourcing of uncertainty

In this section, we've been introduced to two sources of uncertainty, and we've seen how epistemic uncertainty can be very useful for understanding how to interpret our model's outputs. So, you may be wondering: is it possible to separate our sources of uncertainty?

Generally speaking, there are limited guarantees when trying to decompose uncertainty into epistemic and aleatoric components, but some models allow us to obtain a good approximation of this. Ensemble methods provide a particularly good illustrative example.

Let's say we have an ensemble of M models that produce the predictive posterior $P(y|\mathbf{x}, D)$ for some input $\mathbf{x}$ and output y from data D. For a given input, our prediction will have entropy:

$$H[P(y|\mathbf{x}, D)] \approx H\left[\frac{1}{M}\sum_{m=1}^{M} P(y|\mathbf{x}, \theta^m)\right], \theta^m \sim p(\theta|D) \tag{4.6}$$

Here, H denotes entropy, and θ denotes our model parameters. This is a formal expression of concepts we've already covered, showing that the entropy (in other words, uncertainty) of our predictive posterior will be high when our aleatoric and/or epistemic uncertainty is high. This, therefore, represents our **total uncertainty**, which is the uncertainty we'll be working with throughout this book. We can represent this in a manner more consistent with what we'll be encountering in the book – in terms of our predictive standard deviation σ:

$$\sigma = \sigma_a + \sigma_e \tag{4.7}$$

Where a and e denote aleatoric and epistemic uncertainty, respectively.

Because we're working with ensembles, we can go a step further than our total uncertainty. Ensembles are unique in that each model learns something slightly different from the data, due to different data or parameter initialization. As we get an uncertainty estimate for each model, we can take the expectation (in other words, the average) of these uncertainty estimates:

$$\mathbb{E}_{p(\theta|D)}[H[P(y|\mathbf{x}, \theta)]] \approx \frac{1}{M}\sum_{m=1}^{M} H[P(y|\mathbf{x}, \theta^m)], \theta^m \sim p(\theta|D) \tag{4.8}$$

This gives us our **expected data uncertainty** – an estimate of our aleatoric uncertainty. This approximate measure of aleatoric uncertainty becomes more accurate as ensemble size increases. This is possible because of the way ensemble members learn from different subsets of data. If there is no epistemic uncertainty, then the models are consistent, meaning their outputs are identical, and the total uncertainty exclusively comprises the aleatoric uncertainty.

If, on the other hand, there is some epistemic uncertainty, then our total uncertainty comprises both aleatoric and epistemic uncertainty. We can use the expected data uncertainty to determine how much epistemic uncertainty is present in our total uncertainty. We do this using **mutual information**, which is given by:

$$I[y, \theta | \mathbf{x}, D] = H[P(y|\mathbf{x}, D)] - \mathbb{E}_{p(\theta|D)}[H[P(y|\mathbf{x}, \theta)]] \qquad (4.9)$$

We can also express this in terms of equation 4.7:

$$I[y, \theta | \mathbf{x}, D] = \sigma_e = \sigma - \sigma_a \qquad (4.10)$$

As we can see, the concept is pretty straightforward: simply subtract our aleatoric uncertainty from our total uncertainty! The ability to estimate the aleatoric uncertainty can make ensemble methods more attractive for uncertainty quantification, as it allows us to decompose uncertainty, thus providing additional information we don't usually have access to. In *Chapter 6, Bayesian Inference with a Standard Deep Learning Toolbox,* we'll learn more about ensemble techniques for BDL. For non-ensemble methods, we just have the general predictive uncertainty, σ (the combined aleatoric and epistemic uncertainty), which is suitable in most cases.

In the next section, we'll see how we can incorporate uncertainties in how we evaluate our models, and how they can be incorporated in the loss function to improve model training.

4.3.3 Going beyond maximum likelihood: the importance of likelihoods

In the previous section, we saw how uncertainty quantification can help to avoid potentially hazardous scenarios in real-world applications of machine learning. Going back even further to *Chapter 2* and *Chapter 3*, we were introduced to the concept of calibration, and shown how well-calibrated methods' uncertainties increase as data at inference deviates from training data – a concept illustrated in *Figure 4.7*.

While it's easy to illustrate the concept of calibration with simple data – as we saw in *Chapter 2* (through *Figure 2.21*) – unfortunately, it's not easy or practical to do this in most applications. A much more practical approach to understanding how well-calibrated a given method would be to use a metric that incorporates its uncertainty – and this is exactly what we get with **likelihood**.

Likelihood is the probability that some parameters describe some data. As mentioned earlier, we typically work with Gaussian distributions to make things tractable – so we're interested in Gaussian likelihood: the likelihood that the parameters of a Gaussian fit some observed data. The equation for Gaussian likelihood is as follows:

$$p(y) = \frac{1}{\sqrt{2\pi}\sigma} \exp\left\{ -\frac{(y-\mu)^2}{2\sigma^2} \right\} \tag{4.11}$$

Let's see what these distributions would look like for the parameter values we saw earlier in *Figures 4.2* and *4.3*:

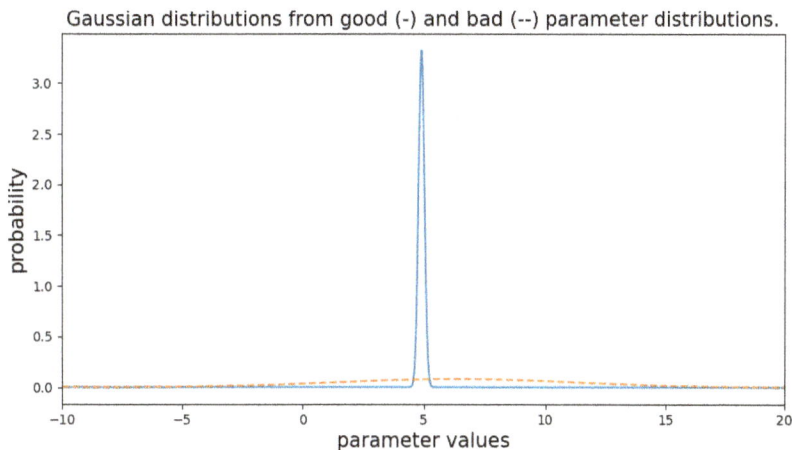

Figure 4.10: The plot of Gaussian distributions corresponding to the parameter sets from Figures 4.2 and 4.3

Visualizing these two distributions highlights the difference in uncertainty between the two parameter sets: our first set of parameters has high probability (solid line), whereas our second set of parameters has low probability (dotted line). But what does this mean for the resulting likelihood values associated with our model's outputs? To investigate these, we need to plug these values into equation 4.11. To do this, we'll need a value for y. We'll use the mean of our target values: 24.03. For our μ and σ values, we'll take the means and standard deviations of the predicted output values, respectively:

$$p(\theta_1) = \frac{1}{\sqrt{2\pi} \times 1.37} \exp\left\{ -\frac{(24.03 - 24.01)^2}{2 \times 1.37^2} \right\} = 0.29 \tag{4.12}$$

$$p(\theta_2) = \frac{1}{\sqrt{2\pi} \times 1.78} \exp\left\{ -\frac{(24.03 - 31.11)^2}{2 \times 1.78^2} \right\} = 7.88 \times 10^{-5} \tag{4.13}$$

We see here that we have a much higher likelihood score for our first set of parameters (θ_1) than for our second (θ_2). This is consistent with *Figure 4.10*, and indicates that, given the data, parameters θ_1 have a higher probability than parameters θ_2 – in other words, parameters θ_1 do a better job of mapping the inputs to the outputs.

These examples illustrate the impact of incorporating uncertainty estimates, allowing us to compute the likelihood of the data. While our error has increased somewhat due to the poorer mean prediction, our likelihood has decreased more dramatically – falling by many orders of magnitude. This tells us that these parameters are doing a very poor job of describing the data, and it does so in a more principled way than simply computing the error between our outputs and our targets.

An important feature of likelihood is that it balances a model's accuracy with its uncertainty. Models that are over-confident have low uncertainty on data for which they have incorrect predictions, and likelihood penalizes them for this overconfidence. Similarly, well-calibrated models are confident on data for which they have correct predictions, and uncertain on data for which they have incorrect predictions. While the models will still be penalized for the incorrect predictions, they will also be rewarded for being uncertain in the right places, and not being over-confident. To see this in practice, we can again use the target output value from the tables shown in *Figure 4.2* and *Figure 4.3*: $y = 24.03$, but we'll also use an incorrect prediction: $y = 5.00$. As we can see, this produces a pretty significant error of $|y - \hat{y}| = |24.03 - 5.00| = 19.03$. Let's take a look at what happens to our likelihood as we increase our σ^2 value associated with this prediction:

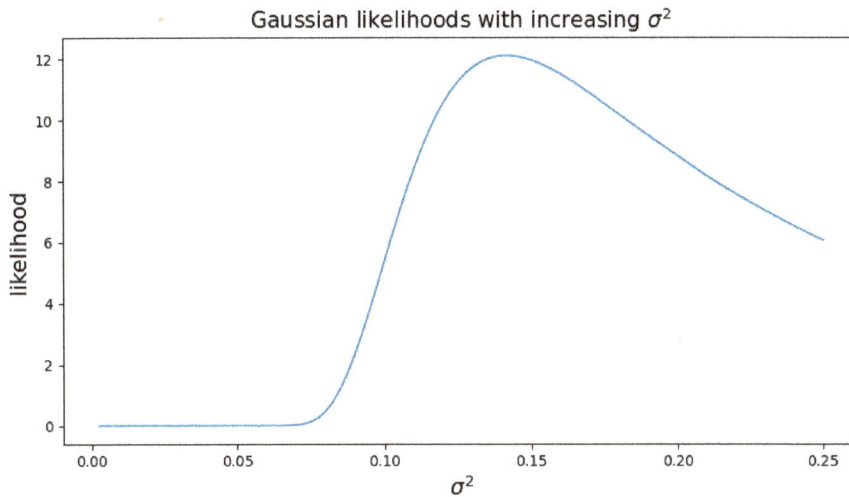

Figure 4.11: A plot of likelihood values with increasing variance

As we see here, our likelihood value is very small when $\sigma^2 = 0.00$, but increases as σ^2 increases to around 0.15, before falling off again. This demonstrates that, given an incorrect prediction, some uncertainty is better than none when it comes to likelihood values. Thus, using likelihoods allows us to train better calibrated models.

Similarly, we can see that if we fix our uncertainty, in this case to $\sigma^2 = 0.1$, and vary our predictions, our likelihood peaks at the correct value, falling off in either direction as our predictions $\hat{y}$ become less accurate and our error $|y - \hat{y}|$ grows:

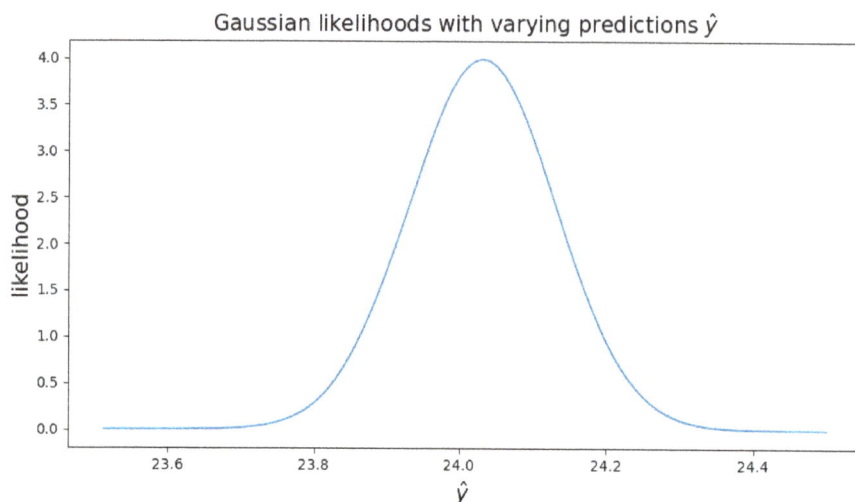

Figure 4.12: A plot of likelihood values with varying predictions

Practically, we don't usually use the likelihood, but instead use the **negative log-likelihood** (**NLL**). We make it negative because, with loss functions, we are interested in finding the minima, rather than the maxima. We use the log because this allows us to use addition, rather than multiplication, which makes things more computationally efficient (making use of the logarithmic identity $log(a * b) = log(a) + log(b)$). The equation that we'll typically be using is therefore:

$$NLL(y) = -\log\left\{\frac{1}{2\pi\sigma}\right\} - \frac{(y-\mu)^2}{2\sigma^2} \tag{4.14}$$

Now that we're familiar with the core concepts of uncertainty and likelihood, we're ready for the next section, where we'll learn how to work with probabilistic concepts in code using the TensorFlow Probability library.

4.4 Tools for BDL

In this chapter, as well as in *Chapter 2*, we've seen a lot of equations involving probability. While it's possible to create BDL models without a probability library, having a library that supports some of the fundamental functions makes things much easier. As we're using TensorFlow for the examples in this book, we'll be using the **TensorFlow Probability (TFP)** library to help us with some of these probabilistic components. In this section, we'll introduce TFP and show how it can be used to easily implement many of the concepts we've seen in *Chapter 2* and *Chapter 4*.

Much of the content up to this point has been about introducing the concept of working with distributions. As such, the first TFP module we'll learn about is the `distributions` module. Let's take a look:

```
1  import tensorflow_probability as tfp
2  tfd = tfp.distributions
3  mu = 0
4  sigma = 1.5
5  gaussian_dist = tfd.Normal(loc=mu, scale=sigma)
```

Here, we have a simple example of initializing a Gaussian (or normal) distribution using the `distributions` module. We can now sample from this distribution – we'll visualize the distribution of our samples using `seaborn` and `matplotlib`:

```
1  import seaborn as sns
2  samples = gaussian_dist.sample(1000)
3  sns.histplot(samples, stat="probability", kde=True)
4  plt.show()
```

This produces the following plot:

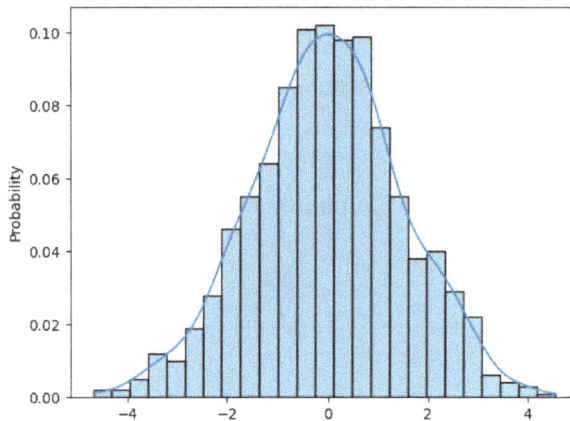

Figure 4.13: A probability distribution of samples drawn from a Gaussian distribution using TFP

As we can see, the samples follow a Gaussian distribution defined by our parameters $\mu = 0$ and $\sigma = 1.5$. The TFD distribution classes also have methods for useful functions such as **Probability Density Function (PDF)** and **Cumulative Density Function (CDF)**. Let's take a look, starting with computing the PDF over a range of values:

```python
pdf_range = np.arange(-4, 4, 0.1)
pdf_values = []
for x in pdf_range:
    pdf_values.append(gaussian_dist.prob(x))
plt.figure(figsize=(10, 5))
plt.plot(pdf_range, pdf_values)
plt.title("Probability density function", fontsize="15")
plt.xlabel("x", fontsize="15")
plt.ylabel("probability", fontsize="15")
plt.show()
```

With the preceding code, we'll produce the following plot:

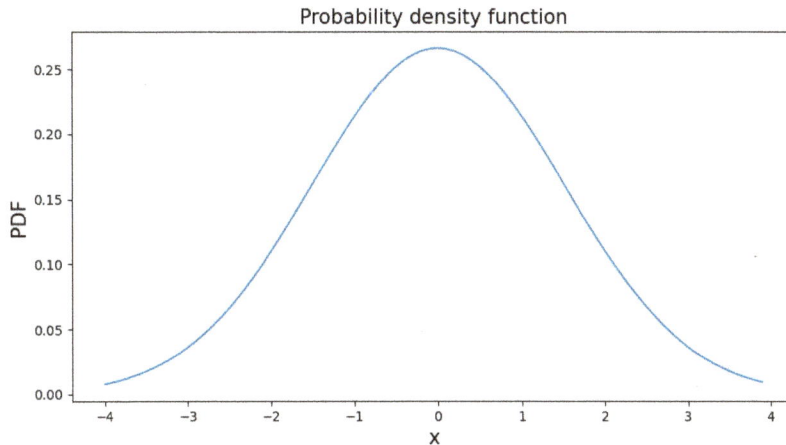

Figure 4.14: A plot of probability density function values for a range of inputs spanning
$x = -4$ *to* $x = 4$

Similarly, we can also compute the CDF:

```
1  cdf_range = np.arange(-4, 4, 0.1)
2  cdf_values = []
3  for x in cdf_range:
4      cdf_values.append(gaussian_dist.cdf(x))
5  plt.figure(figsize=(10, 5))
6  plt.plot(cdf_range, cdf_values)
7  plt.title("Cumulative density function", fontsize="15")
8  plt.xlabel("x", fontsize="15")
9  plt.ylabel("CDF", fontsize="15")
10 plt.show()
```

Compared with the PDF, the CDF produces cumulative probability values, from 0 to 1, as we see in the following plot:

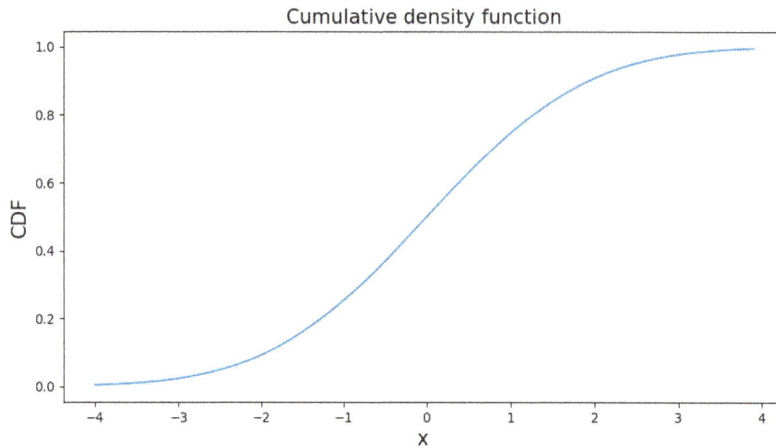

Figure 4.15: Cumulative density function values for a range of inputs spanning $x = -4$ to $x = 4$

The tfp.distributions classes also give us easy access to the parameters of the distributions, for example, we can recover the parameters of our Gaussian distribution via the following:

```
1  mu = gaussian_dist.mean()
2  sigma = gaussian_dist.stddev()
```

Note that these will return tf.Tensor objects, but the NumPy values can be accessed easily via the .numpy() function, for example:

```
1  mu = mu.numpy()
2  sigma = sigma.numpy()
```

This gives us two NumPy scalars for our mu and sigma variables: 0.0 and 1.5, respectively.

Just as we can compute the probability, and thus obtain the PDF, using the prob() function, we can also easily compute the log probability, or log likelihood, using the `log_prob()` function. This makes things a little easier than coding the full likelihood equation (for instance, equation 4.14) each time:

```
1  x = 5
2  log_likelihood = gaussian_dist.log_prob(x)
3  negative_log_likelihood = -log_likelihood
```

Here, we first obtain the log likelihood for some value $x = 5$, and then obtain the NLL, such as would be used in the context of gradient descent.

As we continue through the book, we'll learn more about what TFP has to offer – using the `distributions` module to sample from parameter distributions, and exploring the powerful `tfp.layers` module, which implements probabilistic versions of common neural network layers.

4.5 Summary

In this chapter, we were introduced to the fundamental concepts that we'll need to progress through the book and learn how to implement and use BNNs. Most crucially, we learned about the ideal BNN, which introduced us to the core ideas underlying BDL, and the computational difficulties of achieving this in practice. We also covered the fundamental practical methods used in BDL, giving us a grounding in the concepts that allow us to implement computationally tractable BNNs.

The chapter also introduced the concept of uncertainty sources, describing the difference between data and model uncertainty, how these contribute to total uncertainty, and how we can estimate the contributions of different types of uncertainty with various models. We also introduced one of the most fundamental components in probabilistic inference – the likelihood function – and learned about how it can help us to train better principled and better calibrated models. Lastly, we were introduced to TensorFlow Probability: a powerful library for probabilistic inference, and a crucial component of the practical examples later in the book.

Now that we've covered these fundamentals, we're ready to see how the concepts we've encountered so far can be applied in the implementation of several key BDL models. We'll learn about the advantages and disadvantages of these approaches, and how to apply them to a variety of real-world problems. Continue on to *Chapter 5*, where we'll learn about two key principled approaches for BDL.

4.6   Further reading

This chapter has introduced the material necessary to start working with BDL; however, there are many resources that go into more depth on the topics of uncertainty sources. The following are a few recommendations for readers interested in exploring the theory and code in more depth:

- *Machine Learning: A Probabilistic Perspective, Murphy*: Kevin Murphy's extremely popular book on machine learning has become a staple for students and researchers in the field. This book provides a detailed treatment of machine learning from a probabilistic standpoint, unifying concepts from statistics, machine learning, and Bayesian probability.

- *TensorFlow Probability Tutorials*: in this book, we'll see how TensorFlow Probability can be used to develop BNNs, but their website includes a wide array of tutorials addressing probabilistic programming more generally: `https://www.tensorflow.org/probability/overview`

- *Pyro Tutorials*: Pyro is a PyTorch-based library for probabilistic programming – it's another powerful tool for Bayesian inference, and the Pyro website has many excellent tutorials and examples of probabilistic inference: `https://pyro.ai/`.

5

Principled Approaches for Bayesian Deep Learning

Now that we've introduced the concept of **Bayesian Neural Networks (BNNs)**, we're ready to explore the various ways in which they can be implemented. As we discussed previously, ideal BNNs are computationally intensive, becoming intractable with more sophisticated architectures or larger amounts of data. In recent years, researchers have developed a range of methods that make BNNs tractable, allowing them to be implemented with larger and more sophisticated neural network architectures.

In this chapter, we'll explore two particularly popular methods: **Probabilistic Backpropagation (PBP)** and **Bayes by Backprop (BBB)**. Both methods can be referred to as *probabilistic neural network models*: neural networks designed to learn probabilities over their weights, rather than simply learning point estimates (a fundamental defining feature of BNNs, as we learned in *Chapter 4*). Because they explicitly learn distributions over the weights at training time, we refer to them as *principled* methods; in contrast to the methods we'll explore in the next chapter, which more loosely approximate Bayesian inference with neural networks. We'll cover these topics in the following sections of this chapter:

- Explaining notation

- Familiar probabilistic concepts from deep learning

- Bayesian inference by backpropagation

- Implementing BBB with TensorFlow

- Scalable Bayesian deep learning with PBP

- Implementing PBP

First, let's quickly review the technical requirements for this chapter.

5.1   Technical requirements

To complete the practical tasks in this chapter, you will need a Python 3.8 environment with the Python SciPy stack and the following additional Python packages installed:

- TensorFlow 2.0

- TensorFlow Probability

All of the code for this book can be found on the GitHub repository for the book: `https://github.com/PacktPublishing/Enhancing-Deep-Learning-with-Bayesian-Inference`.

5.2   Explaining notation

While we've introduced much of the notation used throughout the book in the previous chapters, we'll be introducing more notation associated with BDL in the following chapters. As such, we've provided an overview of the notation here for reference:

- μ: The mean. To make it easy to cross-reference our chapter with the original Probabilistic Backpropagation paper, this is represented as m when discussing PBP.

- σ: The standard deviation.

- σ^2: The variance (meaning the square of the standard deviation). To make it easy to cross-reference our chapter with the paper, this is represented as v when discussing PBP.

- **x**: A single vector input to our model. If considering multiple inputs, we'll use **X** to represent a matrix comprising multiple vector inputs.

- $\hat{\mathbf{x}}$: An approximation of our input **x**.

- y: A single scalar target. When considering multiple targets, we'll use **y** to represent a vector of multiple scalar targets.

- $\hat{y}$: A single scalar output from our model. When considering multiple outputs, we'll use $\hat{\mathbf{y}}$ to represent a vector of multiple scalar outputs.

- **z**: The output of an intermediate layer of our model.

- P: Some ideal or target distribution.

- Q: An approximate distribution.

- $KL[Q\|P]$: The KL divergence between our target distribution P and our approximate distribution Q.

- $\mathcal{L}$: The loss.

- $\mathbb{E}$: The expectation.

- $N(\mu, \sigma)$: A normal (or Gaussian) distribution parameterized by the mean μ and the standard deviation σ.

- θ: A set of model parameters.

- Δ: A gradient.

- ∂: A partial derivative.

- $f()$: Some function (e.g. $y = f(x)$ indicates that y is produced by applying function $f()$ to input x).

We will encounter different variations of this notation, using different subscripts or variable combinations.

5.3 Familiar probabilistic concepts from deep learning

While this book introduces many concepts that may be unfamiliar, you may find that some ideas discussed here are familiar. In particular, **Variational Inference (VI)** is something you may be familiar with due to its use in **Variational Autoencoders (VAEs)**.

As a quick refresher, VAEs are generative models that learn encodings that can be used to generate plausible data. Much like standard autoencoders, VAEs comprise an encoder-decoder architecture.

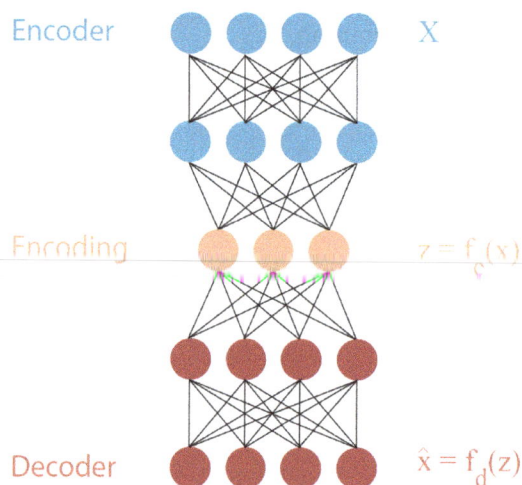

Figure 5.1: Illustration of autoencoder architecture

With a standard autoencoder, the model learns a mapping from the encoder to the latent space, and then from the latent space to the decoder.

As we see here, our output is simply defined as $\hat{x} = f_d(z)$, where our encoding z is simply: $z = f_e(x)$, where $f_e()$ and $f_d()$ are our encoder and decoder functions, respectively. If we want to generate new data using values in our latent space, we could simply inject some random values into the input of our decoder; bypassing the encoder and randomly sampling from our latent space:

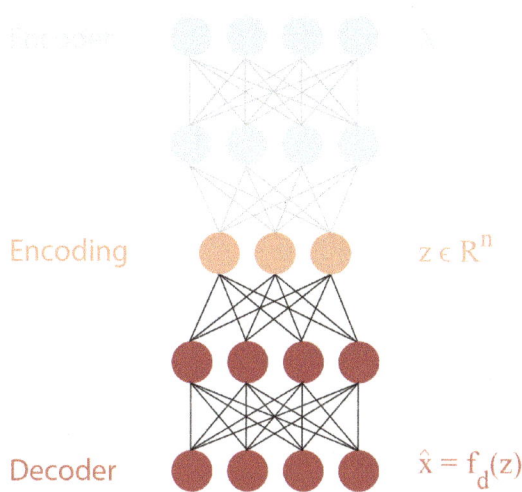

Encoding $z \in R^n$

Decoder $\hat{x} = f_d(z)$

Figure 5.2: Illustration of sampling from the latent space of a standard autoencoder

The problem with this approach is that a standard autoencoder doesn't do a great job of learning the structure of the latent space. This means that while we're free to randomly sample points in this space, there's no guarantee that those points will correspond to something that can be processed by the decoder to generate plausible data.

In a VAE, the latent space is modeled as a distribution. Therefore, what was $z = f_e(x)$ becomes $z \approx \mathcal{N}(\mu_x, \sigma_x)$; that is to say, our latent space z now becomes a Gaussian distribution conditioned on our input x. Now, when we want to generate data using our trained network, we can do so simply by sampling from a normal distribution.

To achieve this, we need to ensure that the latent space approximates a Gaussian distribution. To do so, we use the **Kullback-Leibler divergence** (or KL divergence) during training by incorporating it as a regularization term:

$$\mathcal{L} = \|\mathbf{x} - \hat{\mathbf{x}}\|^2 + KL[Q\|P] \tag{5.1}$$

Here, P is our target distribution (in this case, a multivariate Gaussian distribution), which we're trying to approximate with Q, which is the distribution associated with our latent space, which in this case is as follows:

$$Q = \mathbf{z} \approx \mathcal{N}(\mu, \sigma) \tag{5.2}$$

So, our loss now becomes:

$$\mathcal{L} = \|\mathbf{x} - \hat{\mathbf{x}}\|^2 + KL[q(\mathbf{z}|\mathbf{x})\|p(\mathbf{z})] \tag{5.3}$$

We can expand it as follows:

$$\mathcal{L} = \|\mathbf{x} - \hat{\mathbf{x}}\|^2 + KL[\mathcal{N}(\mu, \sigma)\|\mathcal{N}(0, I)] \tag{5.4}$$

Here, I is the identity matrix. This will allow our latent space to converge on our Gaussian prior, while also minimizing the reconstruction loss. The KL divergence can additionally be rewritten as follows:

$$KL[q(\mathbf{z}|\mathbf{x})\|p(\mathbf{z})] = \underset{q(\mathbf{z}|\mathbf{x})}{\mathbb{E}} \log q(\mathbf{z}|\mathbf{x}) - \underset{q(\mathbf{z}|\mathbf{x})}{\mathbb{E}} \log p(\mathbf{z}) \tag{5.5}$$

The terms on the right hand side of our equation here are the expectation (or mean) of $\log q(\mathbf{z}|\mathbf{x})$ and $\log p(\mathbf{z})$. As we know from *Chapter 2* and *Chapter 4*, we can obtain the the expectation of a given distribution by sampling. Thus, as we can see that all terms of our KL divergence are expectations computed with respect to our approximate distribution $q(\mathbf{z}|\mathbf{x})$, we can approximate our KL divergence by sampling from $q(\mathbf{z}|\mathbf{x})$, which is exactly what we're about to do!

Now that our encoding is represented by the distribution shown in equation 5.2, our neural network structure has to change. We need to learn the mean (μ) and standard deviation (σ) parameters of our distribution:

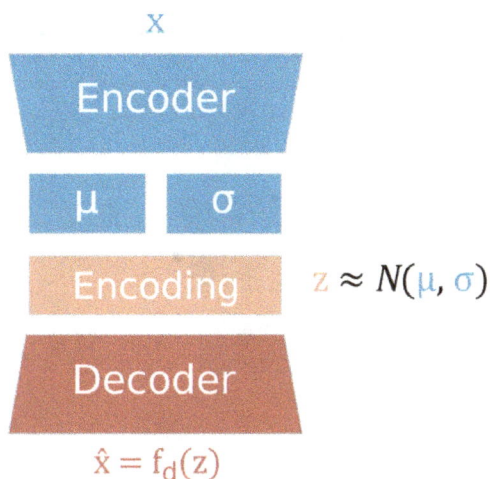

Figure 5.3: Illustration of autoencoder architecture with mean and standard deviation weights

The issue with constructing a VAE in this way is that our encoding z is now stochastic, rather than deterministic. This is a problem because we can't obtain a gradient for stochastic variables – and if we can't obtain a gradient, we have nothing to backpropagate – so we can't learn!

We can fix this using something called the **reparameterization trick**. The reparameterization trick involves modifying how we compute **z**. Instead of sampling **z** from our distribution parameters, we will define it as follows:

$$\mathbf{z} = \mu + \sigma \odot \epsilon \qquad (5.6)$$

As you can see, we've introduced a new variable, ϵ, which is sampled from a Gaussian distribution with $\mu = 0$ and $\sigma = 1$:

$$\epsilon = \mathcal{N}(0, 1) \tag{5.7}$$

Introducing ϵ has allowed us to move the stochasticity out of our backpropagation path. With the stochasticity residing solely in ϵ, we're able to backpropagate through our weights as normal:

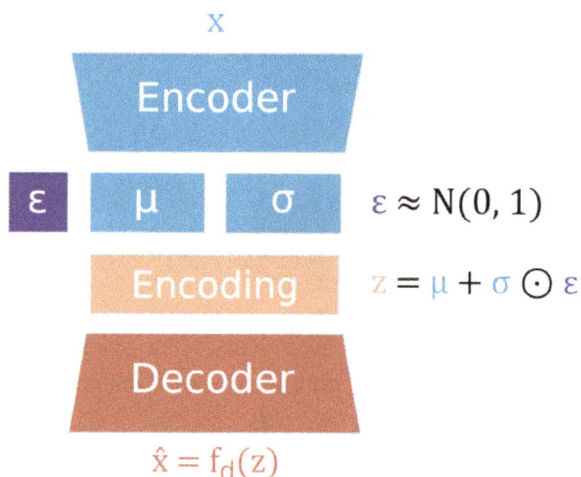

Figure 5.4: Illustration of typical VAE architecture with mean and standard deviation weights, having moved the sampling component out of the backpropagation path

This means we're able to represent our encoding as a distribution while still being able to backpropagate the gradient of z: learning the parameters μ and σ, and using ϵ to sample from the distribution. Being able to represent z as a distribution means we're able to use it to compute the KL divergence, allowing us to incorporate our regularization term in equation 5.1, which in turn allows our embedding to converge towards a Gaussian distribution during training.

These are the fundamental steps in variational learning, and are what turn our standard autoencoder into a VAE. But this isn't all about learning. Crucially for VAEs, because we've learned a normally distributed latent space, we can now sample effectively from that latent space, enabling us to use our VAE to generate new data according to the data landscape learned during training. Unlike the brittle random sampling we had with a standard autoencoder, our VAE is now able to generate *plausible* data!

To do this, we sample ϵ from a normal distribution and multiply σ by this value. This gives us a sample of z to pass through our decoder, obtaining our generated data, $\hat{x}$, at the output.

Now that we're familiar with the fundamentals of variational learning, in the next section we'll see how these principles can be applied to create BNNs.

5.4 Bayesian inference by backpropagation

In their 2015 paper, *Weight Uncertainty in Neural Networks*, Charles Blundell and his colleagues at DeepMind introduced a method for using variational learning for Bayesian inference with neural networks. Their method, which learned the BNN parameters via standard backpropagation, was appropriately named **Bayes by Backprop (BBB)**.

In the previous section, we saw how we can use variational learning to estimate the posterior distribution of our encoding, z, learning $P(z|x)$. For BBB, we're going to be doing very much the same thing, except this time it's not just the encoding we care about. This time we want to learn the posterior distribution over all of the parameters (or weights) of our network: $P(\theta|D)$.

You can think of this as having an entire network made up of VAE encoding layers, looking something like this:

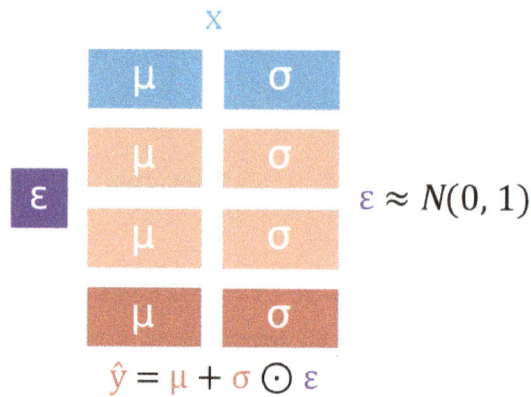

X

$\varepsilon \approx N(0, 1)$

$\hat{y} = \mu + \sigma \odot \varepsilon$

Figure 5.5: Illustration of BBB

As such, it's logical that the learning strategy is also similar to that which we used for the VAE. We again use the principle of variational learning to learn parameters for Q, and approximation of the true distribution P, but this time we're looking for the parameters $\theta^\star$ that minimize this:

$$\theta^\star = \arg\min_{\theta}\ KL[q(\mathbf{w}|\theta)||P(\mathbf{w}|D)] \tag{5.8}$$

Here, D is our data, $\mathbf{w}$ is our network weights, and θ is the parameters of our distribution, e.g. μ and σ in the case of a Gaussian distribution. To do this we make use of an important cost function in Bayesian learning: the **Evidence Lower Bound**[1], or **ELBO** (also referred to as the variational free energy). We denote this with the following:

$$\mathcal{L}(D, \theta) = KL[q(\mathbf{w}|\theta)||P(\mathbf{w})] - \mathop{\mathbb{E}}_{q(\mathbf{w}|\theta)}[\log P(D|\mathbf{w})] \tag{5.9}$$

---

[1]It is beyond the scope of this book to guide the reader through the derivation of ELBO, but we encourage the reader to see the *Further reading* section for texts that provide a more comprehensive overview of ELBO.

This looks rather complicated, but it's really just a generalization of what we saw in equation 5.4. We can break it down as follows:

1. On the left-hand side, we have the KL divergence between our prior $P(\mathbf{w})$ and our approximate distribution $q(\mathbf{w}|\theta)$. This is similar to what we saw in equations 5.1-5.4 in the previous section. Incorporating the KL divergence in our loss allows us to tune our parameters θ such that our approximate distribution converges on our prior distribution.

2. On the right-hand side, we have the expectation of the negative log-likelihood of our data D given our neural network weights $\mathbf{w}$ with respect to the variational distribution. Minimizing this (because it's the *negative log-likelihood*) ensures that we learn parameters that maximize the likelihood of our data given our weights; our network learns to map our inputs to our outputs.

Just as with VAEs, BBB makes use of the reparameterization trick to allow us to backpropagate gradients through our network parameters. Also as before, we sample from our distribution. Taking the form of the KL divergence introduced in equation 5.5, our loss becomes as follows:

$$\mathcal{L}(D, \theta) \approx \sum_{i=1}^{N} \log q(\mathbf{w}_i|\theta) - \log P(\mathbf{w}_i) - \log P(D|\mathbf{w}_i) \qquad (5.10)$$

N is the number of samples, and i denotes a particular sample. While we'll be using Gaussian priors here, an interesting property of this approach is that it can be applied to a wide range of distributions.

The next step is to use our weight samples to train our network:

1. First, just as with VAEs, we sample ϵ from a Gaussian distribution:

$$\epsilon \approx \mathcal{N}(0, I) \qquad (5.11)$$

2. Next, we apply ϵ to the weights in a particular layer, just as with our VAE encoding:

$$\mathbf{w} = \mu + \log(1 + \exp(\rho)) \odot \epsilon \tag{5.12}$$

Note that in BBB, σ is parameterized as $\sigma = \log(1 + \exp(\rho))$. This ensures that it is always non-negative (because a standard deviation cannot be negative!).

3. With our parameters $\theta = (\mu, \rho)$, we define our loss, following equation 3.10, as follows:

$$f(\mathbf{w}, \theta) = \log q(\mathbf{w}|\theta) - \log P(\mathbf{w})P(D|\mathbf{w}) \tag{5.13}$$

4. Because our neural network is made up of weights for both means and standard deviations, we need to calculate the gradients for them separately. We first calculate the gradient with respect to the mean, μ:

$$\Delta_\mu = \frac{\partial f(\mathbf{w}, \theta)}{\partial \mathbf{w}} + \frac{\partial f(\mathbf{w}, \theta)}{\partial \mu} \tag{5.14}$$

Then we calculate the gradient with respect to the standard deviation parameter, ρ:

$$\Delta_\rho = \frac{\partial f(\mathbf{w}, \theta)}{\partial \mathbf{w}} \frac{\epsilon}{1 + \exp(-\rho)} + \frac{\partial f(\mathbf{w}, \theta)}{\partial \rho} \tag{5.15}$$

5. Now we have all the components necessary to update our weights via backpropagation, in a similar fashion to a typical neural network, except we update our mean and variance weights with their respective gradients:

$$\mu \leftarrow \mu - \alpha \Delta_\mu \tag{5.16}$$

$$\rho \leftarrow \rho - \alpha \Delta_\rho \tag{5.17}$$

You may have noticed that the first terms of the gradient computations in equations 5.14 and 5.15 are the gradients you would compute for backpropagation of a typical neural network; we're simply augmenting these gradients with μ- and ρ-specific update rules.

While that was fairly heavy in terms of mathematical content, we can break it down into a few simple concepts:

1. Just as with the encoding in VAEs, we are working with weights that represent the mean and standard deviation of a multivariate distribution, except this time they make up our entire network, not just the encoding layer.

2. Because of this, we again use a loss that incorporates the KL divergence: we're looking to maximize the ELBO.

3. As we are dealing with mean and standard deviation weights, we update these separately with update rules that use the gradients for the respective set of weights.

Now that we understand the core principles behind BBB, we're ready to see how it all comes together in code!

5.5 Implementing BBB with TensorFlow

In this section, we'll see how to implement BBB in TensorFlow. Some of the code you'll see will be familiar; the core concepts of layers, loss functions, and optimizers will be very similar to what we covered in *Chapter 3, Fundamentals of Deep Learning*. Unlike the examples in *Chapter 3*, we'll see how we can create neural networks capable of probabilistic inference.

Step 1: Importing packages

We start by importing the relevant packages. Importantly, we will import `tensorflow-probability`, which will provide us with the layers of the network that replace the point-estimate with a distribution and implement the reparameterization trick. We also set the global parameter for the number of inferences, which will determine how often we sample from the network later:

```
1  import tensorflow as tf
2  import numpy as np
```

```
3  import matplotlib.pyplot as plt
4  import tensorflow_probability as tfp
5
6  NUM_INFERENCES = 7
```

Step 2: Acquiring data

We then download the MNIST Fashion dataset, which is a dataset that contains images of ten different clothing items. We also set the class names and derive the number of training examples and classes:

```
1  # download MNIST fashion data set
2  fashion_mnist = tf.keras.datasets.fashion_mnist
3  (train_images, train_labels), (test_images, test_labels) = fashion_mnist.load_data(
4
5  # set class names
6  CLASS_NAMES = ['T-shirt', 'Trouser', 'Pullover', 'Dress', 'Coat',
7               'Sandal', 'Shirt', 'Sneaker', 'Bag', 'Ankle boot']
8
9  # derive number training examples and classes
10 NUM_TRAIN_EXAMPLES = len(train_images)
11 NUM_CLASSES = len(CLASS_NAMES)
```

Step 3: Helper functions

Next, we create a helper function that defines our model. As you can see, we use a very simple convolutional neural network structure for image classification that consists of a convolutional layer, followed by a max-pooling layer and a fully connected layer. The convolutional layer and the dense layer are imported from the tensorflow-probability package, as indicated by the prefix *tfp*. Instead of defining point-estimates for the weights, they will define weight distributions.

As the names `Convolution2DReparameterization` and `DenseReparameterization` suggest, these layers will use the reparameterization trick to update the weight parameters during backpropagation:

```
1  def define_bayesian_model():
2    # define a function for computing the KL divergence
3    kl_divergence_function = lambda q, p, _: tfp.distributions.kl_divergence(
4      q, p
5    ) / tf.cast(NUM_TRAIN_EXAMPLES, dtype=tf.float32)
6
7    # define our model
8    model = tf.keras.models.Sequential([
9        tfp.layers.Convolution2DReparameterization(
10            64, kernel_size=5, padding='SAME',
11            kernel_divergence_fn=kl_divergence_function,
12            activation=tf.nn.relu),
13        tf.keras.layers.MaxPooling2D(
14            pool_size=[2, 2], strides=[2, 2],
15            padding='SAME'),
16        tf.keras.layers.Flatten(),
17        tfp.layers.DenseReparameterization(
18            NUM_CLASSES, kernel_divergence_fn=kl_divergence_function,
19            activation=tf.nn.softmax)
20    ])
21    return model
```

We also create another helper function that compiles the model for us, using Adam as our optimizer and a categorical cross-entropy loss. Provided with this loss and the preceding network structure, `tensorflow-probability` will automatically add the Kullback-Leibler divergence that is contained in the convolutional and dense layers to the cross-entropy loss. This combination effectively amounts to calculating the ELBO loss that we described in equation 5.9:

```
1  def compile_bayesian_model(model):
2    # define the optimizer
3    optimizer = tf.keras.optimizers.Adam()
4    # compile the model
5    model.compile(optimizer, loss='categorical_crossentropy',
6                  metrics=['accuracy'], experimental_run_tf_function=False)
7    # build the model
8    model.build(input_shape=[None, 28, 28, 1])
9    return model
```

Step 4: model training

Before we can train the model, we first need to convert the labels of the training data from integers to one-hot vectors because this is what TensorFlow expects for the categorical cross-entropy loss. For example, if an image shows a t-shirt and the integer label for t-shirts is 1, then this label will be transformed like this:

[1, 0, 0, 0, 0, 0, 0, 0, 0, 0]:

```
1  train_labels_dense = tf.one_hot(train_labels, NUM_CLASSES)
```

Now, we are ready to train our model on the training data. We will train for ten epochs:

```
1  # use helper function to define the model architecture
2  bayesian_model = define_bayesian_model()
3  # use helper function to compile the model
4  bayesian_model = compile_bayesian_model(bayesian_model)
5  # initiate model training
6  bayesian_model.fit(train_images, train_labels_dense, epochs=10)
```

Step 5: inference

We can then use the trained model to perform inference on the test images. Here, we predict the class label for the first 50 images in the test split. For every image, we sample seven times from the network (as determined by NUM_INFERENCES), which will give us seven predictions for every image:

```
1  NUM_SAMPLES_INFERENCE = 50
2  softmax_predictions = tf.stack(
3      [bayesian_model.predict(test_images[:NUM_SAMPLES_INFERENCE])
4      for _ in range(NUM_INFERENCES)],axis=0)
```

And that's it: we have a working BBB model! Let's visualize the first image in the test split and the seven different predictions for that image. First, we obtain the class predictions:

```
1  # get the class predictions for the first image in the test set
2  image_ind = 0
3  # collect class predictions
4  class_predictions = []
5  for ind in range(NUM_INFERENCES):
6      prediction_this_inference = np.argmax(softmax_predictions[ind][image_ind])
7      class_predictions.append(prediction_this_inference)
8  # get class predictions in human-readable form
9  predicted_classes = [CLASS_NAMES[ind] for ind in class_predictions]
```

Then, we visualize the image along with the predicted class for every inference:

```
1  # define image caption
2  image_caption = []
3  for caption in range(NUM_INFERENCES):
4      image_caption.append(f"Sample {caption+1}: {predicted_classes[caption]}\n")
5  image_caption = ' '.join(image_caption)
6  # visualise image and predictions
7  plt.figure(dpi=300)
8  plt.title(f"Correct class: {CLASS_NAMES[test_labels[image_ind]]}")
```

```
 9   plt.imshow(test_images[image_ind], cmap=plt.cm.binary)
10   plt.xlabel(image_caption)
11   plt.show()
```

Looking at the image in *Figure 5.7*, on most samples the network predicts Ankle boot (which is the correct class). For two of the samples, the network also predicted Sneaker, which is somewhat plausible given the image is showing a shoe.

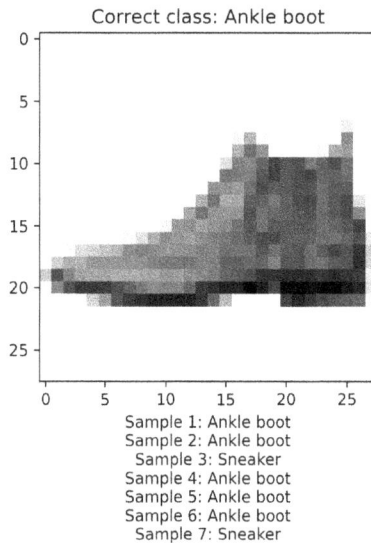

Figure 5.6: Class predictions across the seven different samples from the network trained
with the BBB approach on the first test image in the MNIST fashion dataset

Given that we now have seven predictions per image, we can also calculate the mean variance across these predictions to approximate an uncertainty value:

```
1   # calculate variance across model predictions
2   var_predictions = tf.reduce_mean(
3       tf.math.reduce_variance(softmax_predictions, axis=0),
4       axis=1)
```

For example, the uncertainty value for the first test image in the MNIST fashion dataset is 0.0000002. To put this uncertainty value into context, let's load some images from the regular MNIST dataset, which contains handwritten digits between 0 and 9, and obtain uncertainty values from the model we have trained. We load the dataset and then perform inference again and obtain the uncertainty values:

```
1   # load regular MNIST data set
2   (train_images_mnist, train_labels_mnist),
3   (test_images_mnist, test_labels_mnist) =
4   tf.keras.datasets.mnist.load_data()
5
6   # get model predictions in MNIST data
7   softmax_predictions_mnist =
8       tf.stack([bayesian_model.predict(
9           test_images_mnist[:NUM_SAMPLES_INFERENCE])
10          for _ in range(NUM_INFERENCES)], axis=0)
11
12  # calculate variance across model predictions in MNIST data
13  var_predictions_mnist = tf.reduce_mean(
14      tf.math.reduce_variance(softmax_predictions_mnist, axis=0),
15      axis=1)
```

We can then visualize and compare the uncertainty values between the first 50 images in the fashion MNIST dataset and the regular MNIST dataset.

In *Figure 5.7*, we see that uncertainty values are a lot greater for images from the regular MNIST dataset than for the fashion MNIST dataset. This is expected, given that our model has only seen fashion MNIST images during training and the handwritten digits from the regular MNIST dataset are out-of-distribution for the model we trained.

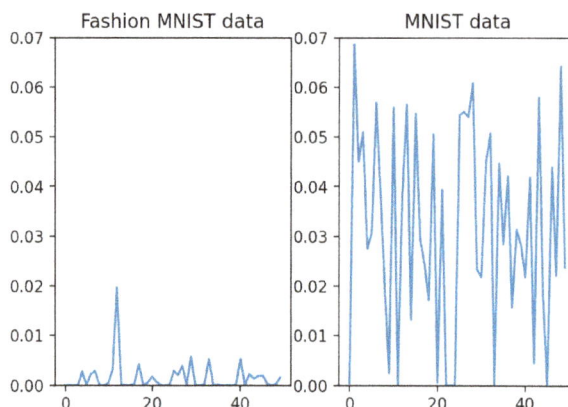

Figure 5.7: Uncertainty values for images in the fashion MNIST dataset (left) versus the regular MNIST dataset (right)

BBB is perhaps the most commonly encountered highly principled Bayesian deep learning method, but it isn't the only option for those concerned with better principled methods. In the next section, we'll introduce another highly principled method and learn about the properties that differentiate it from BBB.

5.6   Scalable Bayesian Deep Learning with Probabilistic Backpropagation

BBB provided a great introduction to Bayesian inference with neural networks, but variational methods have one key drawback: their reliance on sampling at training and inference time. Unlike a standard neural network, we need to sample from the weight parameters using a range of ϵ values in order to produce the distributions necessary for probabilistic training and inference.

At around the same time that BBB was introduced, researchers at Harvard University were working on their own brand of Bayesian inference with neural networks: **Probabilistic Backpropagation**, or **PBP**. Like BBB, PBP's weights form the parameters of a distribution, in this case mean and variance weights (using variance, σ^2, rather than σ). In fact, the similarities don't end here – we're going to see quite a few similarities to

BBB but, crucially, we're going to end up with a different approach to BNN approximation with its own advantages and disadvantages. So, let's get started.

To make things simpler, and for parity with the various PBP papers, we'll stick with individual weights while we work through the core ideas of PBP. Here's a visualization of how these weights are related in a small neural network:

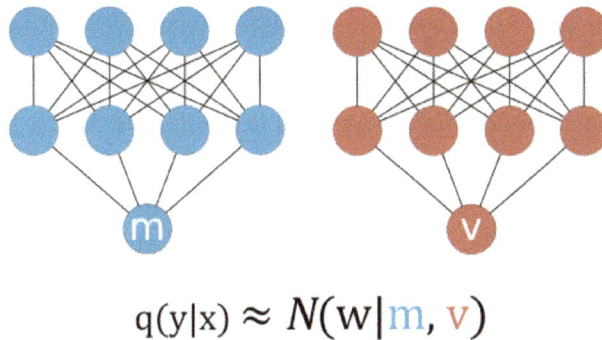

$$q(y|x) \approx N(w|m, v)$$

Figure 5.8: Illustration of neural network weights in PBP

Just as before, we see that our network is essentially built from two sub-networks: one for the mean weights, or m, and one for the variance weights, or v. The core idea behind PBP is that, for each weight, we have some distribution $P(w|D)$ that we're trying to approximate:

$$q(w) = \mathcal{N}(w|m, v) \tag{5.18}$$

This notation should be very familiar now, with $P()$ being the true (intractable) distribution, and $q()$ being the approximate distribution. In PBP's case, as demonstrated in equation 5.18, this is a Gaussian distribution parameterized by mean m and variance v.

In BBB, we saw how variational learning via the ELBO used the KL divergence to ensure our weight distribution converged towards our prior $P(w)$. In PBP, we will again make use of the KL divergence, although this time we'll do it indirectly. The way that we achieve this is through the use of a process called **Assumed Density Filtering (ADF)**.

ADF was developed as a fast sequential method of minimizing the KL divergence between the true posterior $P(w|D)$ and some approximation $q(w|D)$. A key point here is the fact that it is a *sequential* algorithm: just like gradient descent, which we use with standard neural networks, ADF updates its parameters sequentially. This makes it particularly well suited for adapting to a neural network. The ADF algorithm can be described in two key steps:

1. Initialize our parameters, with $m = 0$ and $v = 1$; that is, we start with a unit Gaussian $\mathcal{N}(0, 1)$.

2. Next, we go through each data point $x_i \in \mathbf{x}$ and update the parameters of our model using a set of specific update equations that update our model parameters m and v separately.

While it's beyond the scope of this book to provide a full derivation of ADF, you should know that as we update our parameters through ADF, we also minimize the KL divergence

Thus, for PBP, we need to adapt typical neural network update rules so that the weights are updated along the lines of ADF instead. We do this using the following update rules, which are derived from the original ADF equations:

$$m^{new} = m + v\frac{\partial \log Z}{\partial m} \tag{5.19}$$

$$v^{new} = v - v^2 \left[\left(\frac{\partial \log Z}{\partial m}\right) - 2\frac{\partial \log Z}{\partial v} \right] \tag{5.20}$$

Here, $\log Z$ denotes the Gaussian marginal likelihood, which is defined as follows:

$$\log Z = -\log p(y|m, v) = -0.5 \times \frac{\log v + (y - m)^2}{v} \tag{5.21}$$

This is the **negative log-likelihood (NLL)**. Equation 5.21 is crucial to how we learn the parameters of PBP, as this is the loss function that we're trying to optimise - so let's take some time to understand what's going on. Just as with our loss for BBB (equation 5.9), we can see that our $\log Z$ loss incorporates a few important pieces of information:

1. In the numerator, we see $(y - m)^2$. This is similar to a typical loss we're used to seeing in standard neural network training (the L2 loss). This incorporates the penalty between our target y and our mean estimate for the value, m.

2. The whole equation gives us the NLL function, which describes the joint probability of our target y as a function of our distribution parameterised by m and v.

This has some important properties, which we can explore through a few simple examples. Let's look at the loss for some arbitrary parameters $m = 0.8$ and $v = 0.4$ for a given target $y = 0.6$:

$$-0.5 \times \frac{\log v + (y - m)^2}{v} = -0.5 \times \frac{\log(0.4) + (0.6 - 0.8)^2}{0.4} = 1.095 \tag{5.22}$$

Here, we can see that our typical error, in this case the squared error, is $(0.6 - 0.8)^2 = 0.04$, and we know that as m converges towards y, this will shrink. In addition to that, the log-likelihood scales our error. This is important, because a well-conditioned model for uncertainty quantification will be *more uncertain* when it's wrong, and *more confident* when it's right. The likelihood function gives us a way of achieving this, ensuring that our likelihood is greater if we're uncertain about incorrect predictions and certain about correct predictions.

We can see this in action by substituting another value of v and seeing how this changes the NLL. For example, let's increase our variance to $v = 0.9$:

$$-0.5 \times \frac{\log(0.9) + (0.6 - 0.8)^2}{0.9} = 0.036 \tag{5.23}$$

This significant increase in variance produces a similarly significant reduction in NLL. Similarly, we'll see our NLL increase again if we have high variance for a correct prediction $m = y$:

$$-0.5 \times \frac{\log(0.9) + (0.8 - 0.8)^2}{0.9} = 0.059 \tag{5.24}$$

Hopefully, with this example you can see how using the NLL loss translates to well calibrated uncertainty estimates over our outputs. In fact, this property – using the variance to scale to our objective function – is a fundamental component of all principled BNN methods: BBB also does this, although it's a little more difficult to demonstrate on paper as it requires sampling.

There are a few low-level details of PBP that we'll encounter in the implementation. These relate to the ADF process, and we encourage you to take a look at the articles in the *Further reading* section for comprehensive derivations of PBP and ADF.

Now that we've covered PBP's core concepts, let's take a look at how we implement it with TensorFlow.

5.7 Implementing PBP

Because PBP is quite complex, we'll implement it as a class. Doing so will keep our example code tidy and allow us to easily compartmentalize our various blocks of code. It will also make it easier to experiment with, for example, if you want to explore changing the number of units or layers in your network.

Step 1: Importing libraries

We begin by importing various libraries. In this example, we will use scikit-learn's California Housing dataset to predict house prices:

```
1  from typing import List, Union, Iterable
2  import math
3  from sklearn import datasets
4  from sklearn.model_selection import train_test_split
5  import tensorflow as tf
6  import numpy as np
7  from tensorflow.python.framework import tensor_shape
8  import tensorflow_probability as tfp
```

To make sure we produce the same output every time, we initialize our seeds:

```
1  RANDOM_SEED = 0
2  np.random.seed(RANDOM_SEED)
3  tf.random.set_seed(RANDOM_SEED)
```

We can then load our dataset and create train and test splits:

```
1  # load the California Housing dataset
2  X, y = datasets.fetch_california_housing(return_X_y=True)
3  # split the data (X) and targets (y) into train and test sets
4  X_train, X_test, y_train, y_test = train_test_split(
5      X, y, test_size=0.1, random_state=0
6  )
```

Step 2: Helper functions

Next, we define two helper functions that ensure that our data is in the correct format, one for the input and another one for the output data:

```
1    def ensure_input(x, dtype, input_shape):
2        # a function to ensure that our input is of the correct shape
3        x = tf.constant(x, dtype=dtype)
4        call_rank = tf.rank(tf.constant(0, shape=input_shape, dtype=dtype)) + 1
5        if tf.rank(x) < call_rank:
6            x = tf.reshape(x, [-1, * input_shape.as_list()])
7        return x
8
9
10   def ensure_output(y, dtype, output_dim):
11       # a function to ensure that our output is of the correct shape
12       output_rank = 2
13       y = tf.constant(y, dtype=dtype)
14       if tf.rank(y) < output_rank:
15           y = tf.reshape(y, [-1, output_dim])
16       return y
```

We will also create a short class to initialize a gamma distribution:

ReciprocalGammaInitializer. This distribution is used as the prior for PBP's precision

parameter λ and the noise parameter γ.

```
1    class ReciprocalGammaInitializer:
2        def __init__(self, alpha, beta):
3            self.Gamma = tfp.distributions.Gamma(concentration=alpha, rate=beta)
4
5        def __call__(self, shape: Iterable, dtype=None):
6            g = 1.0 / self.Gamma.sample(shape)
7            if dtype:
8                g = tf.cast(g, dtype=dtype)
9
10           return g
```

A thorough treatment of these variables is not required for a general understanding of PBP. For further details on this, please see the PBP paper listed in the *Further reading* section.

Step 3: Data preparation

With these prerequisites implemented, we can normalize our data. Here, we normalize to mean zero and unit standard deviation. This is a common pre-processing step that will make it easier for our model to find the right set of weights:

```python
1  def get_mean_std_x_y(x, y):
2      # compute the means and standard deviations of our inputs and targets
3      std_X_train = np.std(x, 0)
4      std_X_train[std_X_train == 0] = 1
5      mean_X_train = np.mean(x, 0)
6      std_y_train = np.std(y)
7      if std_y_train == 0.0:
8          std_y_train = 1.0
9      mean_y_train = np.mean(y)
10     return mean_X_train, mean_y_train, std_X_train, std_y_train
11
12 def normalize(x, y, output_shape):
13     # use the means and standard deviations to normalize our inputs and targets
14     x = ensure_input(x, tf.float32, x.shape[1])
15     y = ensure_output(y, tf.float32, output_shape)
16     mean_X_train, mean_y_train, std_X_train, std_y_train = get_mean_std_x_y(x, y)
17     x = (x - np.full(x.shape, mean_X_train)) / np.full(x.shape, std_X_train)
18     y = (y - mean_y_train) / std_y_train
19     return x, y
20
21 # run our normalize() function on our data
22 x, y = normalize(X_train, y_train, 1)
```

Step 4: Defining our model class

We can now start to define our model. Our model will consist of three layers: two ReLU layers and one linear layer. We use Keras' `Layer` to define our layers. The code for this layer is quite long, so we will break it into several subsections.

First, we subclass the `Layer` to create our own `PBPLayer` and define our init method. Our initialization method sets the number of units in our layer:

```
1  from tensorflow.keras.initializers import HeNormal
2
3  # a class to handle our PBP layers
4  class PBPLayer(tf.keras.layers.Layer):
5      def __init__(self, units: int, dtype=tf.float32, *args, **kwargs):
6          super().__init__(dtype=tf.as_dtype(dtype), *args, **kwargs)
7          self.units = units
8      ...
```

We then create a `build()` method that defines the weights of our layer. As we discussed in the previous section, PBP comprises both *mean* weights and *variance* weights. As a simple MLP is composed of a multiplicative component, or weight, and a bias, we'll split both our weights and biases into mean and variance variables:

```
1      ...
2      def build(self, input_shape):
3          input_shape = tensor_shape.TensorShape(input_shape)
4          last_dim = tensor_shape.dimension_value(input_shape[-1])
5          self.input_spec = tf.keras.layers.InputSpec(
6              min_ndim=2, axes={-1: last_dim}
7          )
8          self.inv_sqrtV1 = tf.cast(
9              1.0 / tf.math.sqrt(1.0 * last_dim + 1), dtype=self.dtype
10         )
```

```
11          self.inv_V1 = tf.math.square(self.inv_sqrtV1)
12
13          over_gamma = ReciprocalGammaInitializer(6.0, 6.0)
14          self.weights_m = self.add_weight(
15              "weights_mean", shape=[last_dim, self.units],
16              initializer=HeNormal(), dtype=self.dtype, trainable=True,
17          )
18          self.weights_v = self.add_weight(
19              "weights_variance", shape=[last_dim, self.units],
20              initializer=over_gamma, dtype=self.dtype, trainable=True,
21          )
22          self.bias_m = self.add_weight(
23              "bias_mean", shape=[self.units],
24              initializer=HeNormal(), dtype=self.dtype, trainable=True,
25          )
26          self.bias_v = self.add_weight(
27              "bias_variance", shape=[self.units],
28              initializer=over_gamma, dtype=self.dtype, trainable=True,
29          )
30          self.Normal = tfp.distributions.Normal(
31              loc=tf.constant(0.0, dtype=self.dtype),
32              scale=tf.constant(1.0, dtype=self.dtype),
33          )
34          self.built = True
35      ...
```

The weights_m and weights_v variables are our mean and variance weights, forming the very core of our PBP model. We will continue our definition of PBPLayer when we work through our model fitting function. For now, we can subclass this class to create our ReLU layer:

```
1  class PBPReLULayer(PBPLayer):
2      @tf.function
```

```
3      def call(self, x: tf.Tensor):
4          """Calculate deterministic output"""
5          # x is of shape [batch, prev_units]
6          x = super().call(x)
7          z = tf.maximum(x, tf.zeros_like(x))   # [batch, units]
8          return z
9
10     @tf.function
11     def predict(self, previous_mean: tf.Tensor, previous_variance: tf.Tensor):
12         ma, va = super().predict(previous_mean, previous_variance)
13         mb, vb = get_bias_mean_variance(ma, va, self.Normal)
14         return mb, vb
```

You can see that we overwrite two functions: our `call()` and `predict()` functions. The `call()` function calls our regular linear `call()` function and then applies the ReLU max operation we saw in *Chapter 3*. The `predict()` function calls our regular `predict()` function, but then also calls a new function, `get_bias_mean_variance()`. This function computes the mean and variance of our bias in a numerically stable way, as shown here:

```
1    def get_bias_mean_variance(ma, va, normal):
2        variance_sqrt = tf.math.sqrt(tf.maximum(va, tf.zeros_like(va)))
3        alpha = safe_div(ma, variance_sqrt)
4        alpha_inv = safe_div(tf.constant(1.0, dtype=alpha.dtype), alpha)
5        alpha_cdf = normal.cdf(alpha)
6        gamma = tf.where(
7            alpha < -30,
8            -alpha + alpha_inv * (-1 + 2 * tf.math.square(alpha_inv)),
9            safe_div(normal.prob(-alpha), alpha_cdf),
10           )
11       vp = ma + variance_sqrt * gamma
12       bias_mean = alpha_cdf * vp
13       bias_variance = bias_mean * vp * normal.cdf(-alpha) + alpha_cdf * va * (
```

```
14                 1 - gamma * (gamma + alpha)
15        )
16        return bias_mean, bias_variance
```

With our layer definitions in place, we can build our network. We first create a list of all layers in our network:

```
1   units = [50, 50, 1]
2   layers = []
3   last_shape = X_train.shape[1]
4
5   for unit in units[:-1]:
6       layer = PBPReLULayer(unit)
7       layer.build(last_shape)
8       layers.append(layer)
9       last_shape = unit
10  layer = PBPLayer(units[-1])
11  layer.build(last_shape)
12  layers.append(layer)
```

We then create a `PBP` class that contains the model's `fit()` and `predict()` functions, similar to what you see in a model defined with Keras's `tf.keras.Model` class. Next, we'll see a number of important variables; let's go through them here:

- `alpha` and `beta`: These are parameters of our gamma distribution

- `Gamma`: An instance of the `tfp.distributions.Gamma()` class for our gamma distributions, which is a hyper-prior on PBP's precision parameter λ

- `layers`: This variable specifies the number of layers in the model

- `Normal`: Here, we instantiate an instance of the `tfp.distributions.Normal()` class, which implements a Gaussian probability distribution (in this case, with a mean of 0 and a standard deviation of 1):

```python
1  class PBP:
2      def __init__(
3          self,
4          layers: List[tf.keras.layers.Layer],
5          dtype: Union[tf.dtypes.DType, np.dtype, str] = tf.float32
6      ):
7          self.alpha = tf.Variable(6.0, trainable=True, dtype=dtype)
8          self.beta = tf.Variable(6.0, trainable=True, dtype=dtype)
9          self.layers = layers
10         self.Normal = tfp.distributions.Normal(
11             loc=tf.constant(0.0, dtype=dtype),
12             scale=tf.constant(1.0, dtype=dtype),
13         )
14         self.Gamma = tfp.distributions.Gamma(
15             concentration=self.alpha, rate=self.beta
16         )
17
18     def fit(self, x, y, batch_size: int = 16, n_epochs: int = 1):
19         data = tf.data.Dataset.from_tensor_slices((x, y)).batch(batch_size)
20         for epoch_index in range(n_epochs):
21             print(f"{epoch index=}")
22             for x_batch, y_batch in data:
23                 diff_square, v, v0 = self.update_gradients(x_batch, y_batch)
24                 alpha, beta = update_alpha_beta(
25                     self.alpha, self.beta, diff_square, v, v0
26                 )
27                 self.alpha.assign(alpha)
28                 self.beta.assign(beta)
29
30     @tf.function
31     def predict(self, x: tf.Tensor):
32         m, v = x, tf.zeros_like(x)
```

```
33        for layer in self.layers:
34            m, v = layer.predict(m, v)
35        return m, v
36      ...
```

The PBP class __init__ function creates a number of parameters but essentially initializes our α and β hyper-priors with a normal and a gamma distribution. Furthermore, we save the layers that we created in the previous step.

The fit() function updates the gradients of our layers and then updates our α and β parameters. The function for updating gradients is defined as follows:

```
1      ...
2      @tf.function
3      def update_gradients(self, x, y):
4          trainables = [layer.trainable_weights for layer in self.layers]
5          with tf.GradientTape() as tape:
6              tape.watch(trainables)
7              m, v = self.predict(x)
8              v0 = v + safe_div(self.beta, self.alpha - 1)
9              diff_square = tf.math.square(y - m)
10             logZ0 = logZ(diff_square, v0)
11         grad = tape.gradient(logZ0, trainables)
12         for l, g in zip(self.layers, grad):
13             l.apply_gradient(g)
14         return diff_square, v, v0
```

Before we can update our gradients, we need to propagate them forward through the network. To do so, we'll implement our predict() method:

```
1      # ... PBPLayer continued
2
3      @tf.function
```

```
4      def predict(self, previous_mean: tf.Tensor, previous_variance: tf.Tensor):
5          mean = (
6              tf.tensordot(previous_mean, self.weights_m, axes=[1, 0])
7              + tf.expand_dims(self.bias_m, axis=0)
8          ) * self.inv_sqrtV1
9
10         variance = (
11             tf.tensordot(
12                 previous_variance, tf.math.square(self.weights_m), axes=[1, 0]
13             )
14             + tf.tensordot(
15                 tf.math.square(previous_mean), self.weights_v, axes=[1, 0]
16             )
17             + tf.expand_dims(self.bias_v, axis=0)
18             + tf.tensordot(previous_variance, self.weights_v, axes=[1, 0])
19         ) * self.inv_V1
20
21         return mean, variance
```

Now that we can propagate values through our network, we are ready to implement our loss function. As we saw in the previous section, we use the NLL, which we'll define here:

```
1  pi = tf.math.atan(tf.constant(1.0, dtype=tf.float32)) * 4
2  LOG_INV_SQRT2PI = -0.5 * tf.math.log(2.0 * pi)
3
4
5  @tf.function
6  def logZ(diff_square: tf.Tensor, v: tf.Tensor):
7      v0 = v + 1e-6
8      return tf.reduce_sum(
9          -0.5 * (diff_square / v0) + LOG_INV_SQRT2PI - 0.5 * tf.math.log(v0)
10     )
```

```
11
12
13  @tf.function
14  def logZ1_minus_logZ2(diff_square: tf.Tensor, v1: tf.Tensor, v2: tf.Tensor):
15      return tf.reduce_sum(
16          -0.5 * diff_square * safe_div(v2 - v1, v1 * v2)
17          - 0.5 * tf.math.log(safe_div(v1, v2) + 1e-6)
18      )
```

We can now propagate values through the network and obtain our gradient with respect to our loss (as we would with a standard neural network). This means we're ready to update our gradients by applying the update rules we saw in equations 5.19 and 5.20 for the mean weights and variance weights, respectively:

```
1       # ... PBPLayer continued
2
3       @tf.function
4       def apply_gradient(self, gradient):
5           dlogZ_dwm, dlogZ_dwv, dlogZ_dbm, dlogZ_dbv = gradient
6
7           # Weights
8           self.weights_m.assign_add(self.weights_v * dlogZ_dwm)
9           new_mean_variance = self.weights_v - (
10              tf.math.square(self.weights_v)
11              * (tf.math.square(dlogZ_dwm) - 2 * dlogZ_dwv)
12          )
13          self.weights_v.assign(non_negative_constraint(new_mean_variance))
14
15          # Bias
16          self.bias_m.assign_add(self.bias_v * dlogZ_dbm)
17          new_bias_variance = self.bias_v - (
18              tf.math.square(self.bias_v)
19              * (tf.math.square(dlogZ_dbm) - 2 * dlogZ_dbv)
```

```
20              )
21              self.bias_v.assign(non_negative_constraint(new_bias_variance))
```

As discussed in the previous section, PBP belongs to the class of **Assumed Density Filtering (ADF)** methods. As such, we update the α and β parameters according to ADF's update rules:

```
1   def update_alpha_beta(alpha, beta, diff_square, v, v0):
2       alpha1 = alpha + 1
3       v1 = v + safe_div(beta, alpha)
4       v2 = v + beta / alpha1
5       logZ2_logZ1 = logZ1_minus_logZ2(diff_square, v1=v2, v2=v1)
6       logZ1_logZ0 = logZ1_minus_logZ2(diff_square, v1=v1, v2=v0)
7       logZ_diff = logZ2_logZ1 - logZ1_logZ0
8       Z0Z2_Z1Z1 = safe_exp(logZ_diff)
9       pos_where = safe_exp(logZ2_logZ1) * (alpha1 - safe_exp(-logZ_diff) * alpha)
10      neg_where = safe_exp(logZ1_logZ0) * (Z0Z2_Z1Z1 * alpha1 - alpha)
11      beta_denomi = tf.where(logZ_diff >= 0, pos_where, neg_where)
12      beta = safe_div(beta, tf.maximum(beta_denomi, tf.zeros_like(beta)))
13
14      alpha_denomi = Z0Z2_Z1Z1 * safe_div(alpha1, alpha) - 1.0
15
16      alpha = safe_div(
17          tf.constant(1.0, dtype=alpha_denomi.dtype),
18          tf.maximum(alpha_denomi, tf.zeros_like(alpha)),
19      )
20
21      return alpha, beta
```

Step 5: Avoiding numerical errors

Finally, let's define a few helper functions to ensure that we avoid numerical errors during fitting:

```
1   @tf.function
2   def safe_div(x: tf.Tensor, y: tf.Tensor, eps: tf.Tensor = tf.constant(1e-6)):
3       _eps = tf.cast(eps, dtype=y.dtype)
4       return x / (tf.where(y >= 0, y + _eps, y - _eps))
5
6
7   @tf.function
8   def safe_exp(x: tf.Tensor, BIG: tf.Tensor = tf.constant(20)):
9       return tf.math.exp(tf.math.minimum(x, tf.cast(BIG, dtype=x.dtype)))
10
11
12  @tf.function
13  def non_negative_constraint(x: tf.Tensor):
14      return tf.maximum(x, tf.zeros_like(x))
```

Step 6: Instantiating our model

And there we have it: the core code for training PBP. Now we're ready to instantiate our model and train it on some data. Let's use a small batch size and a single epoch in this example:

```
1   model = PBP(layers)
2   model.fit(x, y, batch_size=1, n_epochs=1)
```

Step 7: Using our model for inference

Now that we have our fitted model, let's see how well it works on our test set. We first normalize our test set:

```
1   # Compute our means and standard deviations
2   mean_X_train, mean_y_train, std_X_train, std_y_train = get_mean_std_x_y(
3       X_train, y_train
4   )
```

```
5
6   # Normalize our inputs
7   X_test = (X_test - np.full(X_test.shape, mean_X_train)) /
8       np.full(X_test.shape, std_X_train)
9
10  # Ensure that our inputs are of the correct shape
11  X_test = ensure_input(X_test, tf.float32, X_test.shape[1])
```

Then we get our model predictions: the mean and variance:

```
1   m, v = model.predict(X_test)
```

Then we post-process these values to make sure they have the right shape and are in the range of the original input data:

```
1   # Compute our variance noise - the baseline variation we observe in our targets
2   v_noise = (model.beta / (model.alpha - 1) * std_y_train**2)
3
4   # Rescale our mean values
5   m = m * std_y_train + mean_y_train
6
7   # Rescale our variance values
8   v = v * std_y_train**2
9
10  # Reshape our variables
11  m = np.squeeze(m.numpy())
12  v = np.squeeze(v.numpy())
13  v_noise = np.squeeze(v_noise.numpy().reshape(-1, 1))
```

Now that we've got our predictions, we can compute how well our model's done. We'll use a standard error metric, RMSE, as well as the metric we used for our loss: the NLL. We can compute them using the following:

```
1  rmse = np.sqrt(np.mean((y_test - m) ** 2))
2  test_log_likelihood = np.mean(
3      -0.5 * np.log(2 * math.pi * v)
4      - 0.5 * (y_test - m) ** 2 / v
5  )
6  test_log_likelihood_with_vnoise = np.mean(
7      -0.5 * np.log(2 * math.pi * (v + v_noise))
8      - 0.5 * (y_test - m) ** 2 / (v + v_noise)
9  )
```

Evaluating both of these metrics is good practice for any regression task for which you have model uncertainty estimates. The RMSE gives you your standard error metric, which allows you to compare directly with non-probabilistic methods. The NLL gives you an impression of how well calibrated your method is by evaluating how confident your model is when it's doing well versus doing poorly, as we discussed earlier in the chapter. Together, these metrics give you a comprehensive impression of a Bayesian model's performance, and you'll see them used time and time again in the literature.

5.8 Summary

In this chapter, we learned about two fundamental, well-principled, Bayesian deep learning models. BBB showed us how we can make use of variational inference to efficiently sample from our weight space and produce output distributions, while PBP demonstrated that it's possible to obtain predictive uncertainties *without* sampling. This makes PBP more computationally efficient than BBB, but each model has its pros and cons.

In BBB's case, while it's less computationally efficient than PBP, it's also more adaptable (particularly with the tools available in TensorFlow for variational layers). We can apply this to a variety of different DNN architectures with relatively little difficulty. The price is incurred through the sampling required at both inference and training time: we need to do more than just a single forward pass to obtain our output distributions.

Conversely, PBP allows us to obtain our uncertainty estimates with a single pass, but – as we've just seen – it's quite complex to implement. This makes it awkward to adapt to other network architectures, and while it has been done (see the *Further reading* section), it's not a particularly practical method to use given the technical overhead of implementation and the relatively marginal gains compared to other methods.

In summary, these methods are excellent if you need robust, well-principled BNN approximations and aren't constrained in terms of memory or computational overheads at inference. But what if you have limited memory and/or limited compute, such as running on edge devices? In these cases, you may want to turn to more practical methods of obtaining predictive uncertainties.

In *Chapter 6, Bayesian Neural Network Approximation Using a Standard Deep Learning Toolbox*, we'll see how we can use more familiar components in TensorFlow to create more practical probabilistic neural network models.

5.9 Further reading

- *Weight Uncertainty in Neural Networks*, Charles Blundell *et al.*: This is the paper that introduced BBB, and is one of the key pieces of BDL literature.

- *Practical Variational Inference for Neural Networks*, Alex Graves *et al.*: An influential paper on the use of variational inference for neural networks, this work introduces a straightforward stochastic variational method that can be applied to a variety of neural network architectures.

- *Probabilistic Backpropagation for Scalable Learning of Bayesian Neural Networks*, José Miguel Hernández-Lobato *et al.*: Another important work in BDL literature, this work introduced PBP, demonstrating how Bayesian inference can be achieved via more scalable means.

- *Practical Considerations for Probabilistic Backpropagation*, Matt Benatan *et al.*: In this work, the authors introduce methods for making PBP more practical for real-world applications.

- *Fully Bayesian Recurrent Neural Networks for Safe Reinforcement Learning*, Matt Benatan *et al.*: This paper shows how PBP can be adapted to an RNN architecture, and shows how BNNs can be advantageous in safety-critical systems.

6

Using the Standard Toolbox for Bayesian Deep Learning

As we saw in previous chapters, vanilla NNs often produce poor uncertainty estimates and tend to make overconfident predictions, and some aren't capable of producing uncertainty estimates at all. By contrast, probabilistic architectures offer principled means to obtain high-quality uncertainty estimates; however, they have a number of limitations when it comes to scaling and adaptability.

While both PBP and BBB can be implemented with popular ML frameworks (as shown in our previous TensorFlow examples), they are very complex. As we saw in the last chapter, implementing even a simple network isn't straightforward. This means that adapting them to new architectures is awkward and time-consuming (particularly for PBP, although it is possible – see *Fully Bayesian Recurrent Neural Networks for Safe Reinforcement Learning*). For simple tasks, such as the examples from *Chapter 5, Principled Approaches for Bayesian Deep Learning*, this isn't an issue. But in many real-world tasks, such as machine translation or object recognition, far more sophisticated network architectures are necessary.

While some academic institutions or large research organizations may have the time and resources required to adapt these complex probabilistic methods to a variety of sophisticated architectures, in many cases this simply is not viable. Additionally, more and more industry researchers and engineers are turning to transfer learning-based methods, using pre-trained networks as the backbone of their models. In these cases, it's impossible to simply add probabilistic machinery to predefined architectures.

To address this, in this chapter, we will explore how common paradigms in deep learning can be harnessed to develop probabilistic models. The methods introduced here show that, with relatively minor tweaks, you can easily adapt large, sophisticated architectures to produce high-quality uncertainty estimates. We'll even introduce techniques that will enable you to get uncertainty estimates from networks you've already trained!

The chapter will cover three key approaches for facilitating model uncertainty estimation easily with common deep learning frameworks. First, we will look at **Monte Carlo Dropout** (**MC dropout**), a method that induces variance across predictions by utilizing dropout at inference time. Second, we will introduce deep ensembles, whereby multiple neural networks are combined to facilitate both uncertainty estimation and improved model performance. Finally, we will explore various methods for adding a Bayesian layer to our model, allowing any model to produce uncertainty estimates.

These topics will be covered in the following sections:

- Introducing approximate Bayesian inference via dropout

- Using ensembles for model uncertainty estimates

- Exploring neural network augmentation with Bayesian last-layer methods

6.1 Technical requirements

To complete the practical tasks in this chapter, you will need a Python 3.8 environment with the SciPy stack and the following additional Python packages installed:

- TensorFlow 2.0

- TensorFlow Probability

All of the code for this book can be found on the GitHub repository for the book:
`https://github.com/PacktPublishing/Enhancing-Deep-Learning-with-Bayesian-Inference`.

6.2 Introducing approximate Bayesian inference via dropout

Dropout is traditionally used to prevent overfitting an NN. First introduced in 2012, it is now used in many common NN architectures and is one of the easiest and most widely used regularization methods. The idea of dropout is to randomly turn off (or drop) certain units of a neural network during training. Because of this, the model cannot solely rely on a particular small subset of neurons to solve the task it was given. Instead, the model is forced to find different ways to solve its task. This improves the robustness of the model and makes it less likely to overfit.

If we simplify a network to $y = Wx$, where y is the output of our network, x the input, and W our model weights, we can think of dropout as:

$$\hat{w}_j = \begin{cases} w_j, & p \\ 0, & \text{otherwise} \end{cases} \qquad (6.1)$$

where $\hat{w}_j$ is the new weights after applying dropout, w_j is our weights before applying dropout, and p is our probability of *not* applying dropout.

The original dropout paper recommends randomly dropping 50% of the units in a network and applying dropout to all layers. Input layers should not have the same dropout probability because this would mean that we throw away 50% of the input information for our network, which makes it more difficult for the model to converge. In practice, you can experiment with different dropout probabilities to find the dropout rate that works well for your specific dataset and model; that is another hyperparameter you can optimize. Dropout is typically available as a standalone layer in all standard neural network libraries you can find online. You typically add it after your activation function:

```python
from tensorflow.keras import Sequential
from tensorflow.keras.layers import Flatten, Conv2D, MaxPooling2D, Dropout, Dense


model = Sequential([
    Conv2D(32, (3,3), activation="relu", input_shape=(28, 28, 1)),
    MaxPooling2D((2,2)),
    Dropout(0.2),
    Conv2D(64, (3,3), activation="relu"),
    MaxPooling2D((2,2)),
    Dropout(0.5),
    Flatten(),
    Dense(64, activation="relu"),
    Dropout(0.5),
    Dense(10)
])

```

Now that we've been reminded of the vanilla application of dropout, let's look at how we can use it for Bayesian inference.

6.2.1 Using dropout for approximate Bayesian inference

Traditional dropout methods make the prediction of dropout networks deterministic at test time by turning off dropout during inference. However, we can also use the stochasticity of dropout to our advantage. This is called **Monte Carlo (MC)** dropout, and the idea is as follows:

1. We use dropout during test time.

2. Instead of running inference once, we run it many times (for example, 30-100 times).

3. We then average the predictions to get our uncertainty estimates.

Why is this beneficial? As we said before, using dropout forces the model to learn different ways to solve its task. So, when we keep dropout enabled during inference, we use slightly different networks that all process the input via a slightly different path through the model. This diversity is helpful when we want a calibrated uncertainty score, as we will see in the next section, where we will discuss the concept of deep ensembles. Instead of predicting a point estimate (a single value) for each input, our network now produces a distribution of values (made up of multiple forward passes). We can use this distribution to compute a mean and variance associated with each input data point, as shown in *Figure 6.1*.

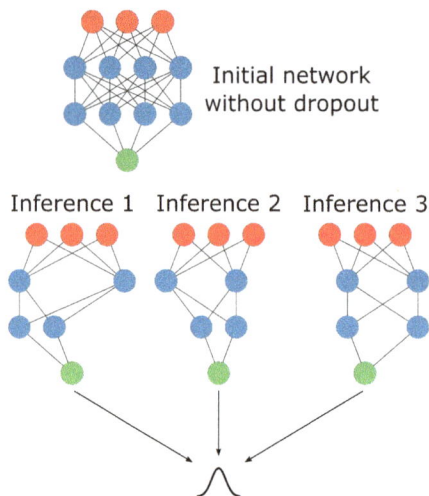

Figure 6.1: Example of MC dropout

We can also interpret MC dropout in a Bayesian way. Using these slightly different networks with dropout can be seen as sampling from a distribution of all possible models: the posterior distribution over all of the parameters (or weights) of our network:

$$\theta_t \sim P(\theta|D) \tag{6.2}$$

Here, θ_t is a dropout configuration and $\sim$ a single sample from our posterior distribution $P(\theta|D)$. This way, MC dropout is equivalent to a form of approximate Bayesian inference, similar to the methods we saw in *Chapter 5*.

Now that we have an idea of how MC dropout works, let's implement it in TensorFlow.

6.2.2 Implementing MC dropout

Let's assume we have trained a model with the convolutional architecture described in this chapter's first hands-on exercise. We can now use dropout at inference by setting `training=True`:

```
1  def mc_dropout_inference(
2      imgs: np.ndarray,
3      nb_inference: int,
4      model: Sequential
5  ) -> np.ndarray:
6      """
7      Run inference nb_inference times with random dropout enabled
8      (training=True)
9      """
10     preds = []
11     for _ in range(nb_inference):
12         preds.append(model(imgs, training=True))
13     return tf.nn.softmax(preds, axis=-1).numpy()
14
15
16 predictions = mc_dropout_inference(test_images, 50, model)
```

This allows us to get our mean and variance for every prediction of our model. Each row of our `predictions` variable contains the predictions associated with each input, obtained from consecutive forward passes. From these predictions, we can compute the means and variances, as follows:

```
1  predictive_mean = np.mean(predictions, axis=0)
2  predictive_variance = np.var(predictions, axis=0)
```

As with all neural networks, Bayesian neural networks require some degree of fine-tuning via hyperparameters. The following three hyperparameters are particularly important for MC dropout:

- **Number of dropout layers**: How many layers (in our `Sequential` object) will use dropout, and which layers these will be.

- **Dropout rate**: The likelihood that nodes will be dropped.

- **The number of MC dropout samples**: A new hyperparameter specific to MC dropout. Shown here as nb_inference, this defines the number of times to sample from the MC dropout network at inference time.

We've now seen how MC dropout can be used in a new way, giving us an easy and intuitive method to compute Bayesian uncertainties using familiar tools. But this isn't the only method we have available to us. In the next section, we'll see how ensembling can be applied to neural networks; providing another straightforward approach for approximating BNNs.

6.3   Using ensembles for model uncertainty estimates

This section will introduce you to deep ensembles: a popular method for obtaining Bayesian uncertainty estimates using an ensemble of deep networks.

6.3.1   Introducing ensembling methods

A common strategy in machine learning is to combine several single models into a committee of models. The process of learning such a combination of models is called **ensemble learning**, and the resulting committee of models is called an **ensemble**. Ensemble learning involves two main components: first, the different single models need to be trained. There are various strategies to obtain different models from the same training data: the models can be trained on different subsets of data, we can train different model types or models with different architectures, or we can initialize the same model types with different hyperparameters. Second, the outputs of the different single models need to be combined. Common strategies for combining the predictions of single models are simply taking their average or taking a majority vote among all members of the ensemble. More advanced strategies are taking a weighted average or, if more training data is available, learning an additional model to combine the different predictions of the ensemble members.

Ensembles are very popular in machine learning because they tend to improve predictive performance by minimizing the risk of accidentally picking a model with poor performance. In fact, ensembles are guaranteed to perform at least as well as any single model. Furthermore, ensembles will perform better than single models if there is enough diversity among the predictions of ensemble members. Diversity here means that different ensemble members make different mistakes on a given data sample. If, for example, some ensemble members misclassify the image of a dog as "cat," but the majority of ensemble members make the correct prediction ("dog"), then the combined ensemble output will still be correct ("dog"). More generally, as long as every single model has an accuracy greater than 50% and the models make independent mistakes, then the predictive performance of the ensemble will approach 100% accuracy as we add more and more ensemble members.

In addition to improving predictive performance, we can leverage the degree of agreement (or disagreement) among ensemble members to obtain an uncertainty estimate along with the prediction of the ensemble. In the context of image classification, for example, if almost all ensemble members predict that the image shows a dog, then we can say that the ensemble predicted "dog" with high confidence (or low uncertainty). Conversely, if there is significant disagreement among the predictions of different ensemble members, then we will observe high uncertainty in the form of significant variance across the outputs from the ensemble members, telling us that the prediction has low confidence.

Now that we are equipped with a basic understanding of ensembles, it is worth highlighting that MC dropout, which we explored in the previous section, may also be seen as an ensemble method. When we enable dropout during inference, we effectively run inference with a slightly different (sub-)network every time. The combination of these different sub-networks can be considered as a committee of different models, and therefore an ensemble. This observation led a team at Google to look into alternative ways of creating ensembles from DNNs, which led to the discovery of deep ensembles (Lakshminarayan et al, 2016) , which are introduced in the following section.

6.3.2 Introducing deep ensembles

The main idea behind deep ensembles is straightforward: train several different DNN models, then combine their predictions via averaging to improve model performance and leverage the agreement among the predictions of these models to obtain an estimate of predictive uncertainty.

More formally, assume that we have some training data $\mathbf{X}$, where $X \in \mathbb{R}^D$, and corresponding target labels $\mathbf{y}$. For example, in image classification the training data would be images, and the target labels would be integers that denote which object class is shown in the corresponding image, so $y \in \{1, ..., K\}$ where K is the total number of classes. Training a single neural network means that we model the probabilistic predictive distribution $p_\theta(y|x)$ over the labels and optimize θ, the parameters of the NN. For deep ensembles, we train $\mathbf{M}$ neural networks whose parameters can be described as $\{\theta_m\}_{m=1}^{M}$, where each θ_m is optimized independently using $\mathbf{X}$ and $\mathbf{y}$ (meaning that we train each NN independently on the same data). The predictions of the deep ensemble members are combined via averaging, using $p(y|x) = M^{-1} \sum_{m=1}^{M} p_{\theta_m}(y|x, \theta_m)$.

Figure 6.2 illustrates the idea behind deep ensembles. Here, we have trained $M = 3$ different feed-forward NNs. Notice that each network has its own unique set of network weights, as illustrated by the varying thickness of the edges connecting the network notes. Each of the three networks will output its own prediction score, as illustrated by the green nodes, and we combine these scores via averaging.

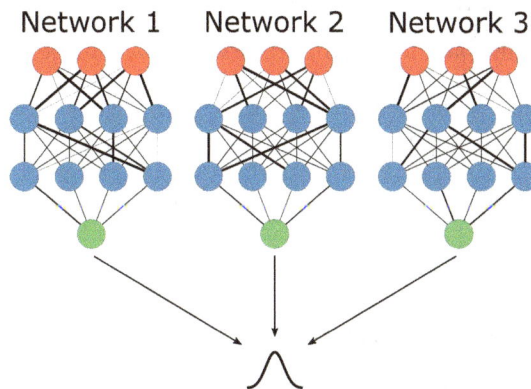

Figure 6.2: Example of a deep ensemble. Note that the three networks differ in their weights, as illustrated by edges with different thicknesses

How can we train several different neural network models if only one dataset is available for training? The strategy proposed in the original paper (and still the most commonly used strategy) is to start every training with a random initialization of the network's weights. If every training starts with a different set of weights, the different training runs are likely to produce networks with different function approximations of the training data. This is because NNs tend to have many more weight parameters than there are samples in the training dataset. Therefore, the same observations in the training dataset can be approximated by many different weight parameter combinations. During training, the different NN models will each converge to their own parameter combination and will occupy different local optima on the loss landscape. Because of this, the different NNs will also often have differing perspectives on a given data sample, for example, the image of a dog. This also means that the different NNs will make different mistakes, for example, when classifying the data sample. The degree of consensus between the different networks in an ensemble provides information about how certain an ensemble is in its predictions for a given data point: the more the networks agree, the more confident we can be in the prediction.

Alternative ways to train different NN models with the same training data set are: to use a random ordering of mini-batches during training, to use different hyperparameters for every training run, or to use different network architecture for every model altogether. These strategies can also be combined, and understanding exactly which combination of strategies leads to the best outcomes, in terms predictive performance and predictive uncertainty, is an active field of research.

6.3.3   Implementing a deep ensemble

The following code example illustrates how to train a deep ensemble using the strategy of random weight initialization to obtain differing ensemble members.

Step 1: Importing libraries

We start by importing the relevant packages and setting the number of ensembles to 3 for this code example:

```
1  import tensorflow as tf
2  import numpy as np
3  import matplotlib.pyplot as plt
4
5
6  ENSEMBLE_MEMBERS = 3
```

Step 2: Obtaining data

We then download the MNIST Fashion dataset, which is a dataset that contains images of ten different clothing items:

```
1  # download data set
2  fashion_mnist = tf.keras.datasets.fashion_mnist
3  # split in train and test, images and labels
4  (train_images, train_labels), (test_images, test_labels) = fashion_mnist.load_data(
5
```

```
6  # set class names
7  CLASS_NAMES = ['T-shirt', 'Trouser', 'Pullover', 'Dress', 'Coat',
8                  'Sandal', 'Shirt', 'Sneaker', 'Bag', 'Ankle boot']
```

Step 3: Constructing our ensemble model

Next, we create a helper function that defines our model. As you can see, we use a simple image classifier structure that consists of two convolutional layers, each followed by a max-pooling operation, and several fully connected layers:

```
1  def build_model():
2    # we build a forward neural network with tf.keras.Sequential
3    model = tf.keras.Sequential([
4        # we define two convolutional layers followed by a max-pooling operation each
5        tf.keras.layers.Conv2D(filters=32, kernel_size=(5,5), padding='same',
6                                activation='relu', input_shape=(28, 28, 1)),
7        tf.keras.layers.MaxPool2D(strides=2),
8        tf.keras.layers.Conv2D(filters=48, kernel_size=(5,5), padding='valid',
9                                activation='relu'),
10       tf.keras.layers.MaxPool2D(strides=2),
11       # we flatten the matrix output into a vector
12       tf.keras.layers.Flatten(),
13       # we apply three fully-connected layers
14       tf.keras.layers.Dense(256, activation='relu'),
15       tf.keras.layers.Dense(84, activation='relu'),
16       tf.keras.layers.Dense(10)
17   ])
18
19   return model
20
```

We also create another helper function that compiles the model for us, using Adam as our optimizer and a categorical cross-entropy loss:

```
1  def compile_model(model):
2    model.compile(optimizer='adam',
3                  loss=tf.keras.losses.SparseCategoricalCrossentropy(from_logits=True)
4                  metrics=['accuracy'])
5    return model
6
```

Step 4: Training

We then train three different networks on the same dataset. Since the network weights
are initialized at random, this will result in three different models. You will see that the
training accuracy varies slightly between models:

```
1  deep_ensemble = []
2  for ind in range(ENSEMBLE_MEMBERS):
3      model = build_model()
4      model = compile_model(model)
5      print(f"Train model {ind:02}")
6      model.fit(train_images, train_labels, epochs=10)
7      deep_ensemble.append(model)
```

Step 5: Inference

We can then perform inference and obtain the predictions for each of the models for all
images in the test split. We can also take the mean across the predictions of the three
models, which will give us one prediction vector per image:

```
1  # get logit predictions for all three models for images in the test split
2  ensemble_logit_predictions = [model(test_images) for model in deep_ensemble]
3  # convert logit predictions to softmax
4  ensemble_softmax_predictions = [
5      tf.nn.softmax(logits, axis=-1) for logits in ensemble_logit_predictions]
6
```

```
7  # take mean across models, this will result in one prediction vector per image
8  ensemble_predictions = tf.reduce_mean(ensemble_softmax_predictions, axis=0)
```

That's it. We have trained an ensemble of networks and performed inference. Given that we have several predictions per image now, we can also look at images where the three models disagree.

Let's, for example, find the image with the highest disagreement and visualize it:

```
1   # calculate variance across model predictions
2   ensemble_std = tf.reduce_mean(
3       tf.math.reduce_variance(ensemble_softmax_predictions, axis=0),
4       axis=1)
5   # find index of test image with highest variance across predictions
6   ind_disagreement = np.argmax(ensemble_std)
7
8   # get predictions per model for test image with highest variance
9   ensemble_disagreement = []
10  for ind in range(ENSEMBLE_MEMBERS):
11      model_prediction = np.argmax(ensemble_softmax_predictions[ind][ind_disagreement])
12      ensemble_disagreement.append(model_prediction)
13  # get class predictions
14  predicted_classes = [CLASS_NAMES[ind] for ind in ensemble_disagreement]
15
16  # define image caption
17  image_caption = \
18      f"Network 1: {predicted_classes[0]}\n" + \
19      f"Network 2: {predicted_classes[1]}\n" + \
20      f"Network 3: {predicted_classes[2]}\n"
21
22  # visualise image and predictions
23  plt.figure()
24  plt.title(f"Correct class: {CLASS_NAMES[test_labels[ind_disagreement]]}")
```

```
25  plt.imshow(test_images[ind_disagreement], cmap=plt.cm.binary)
26  plt.xlabel(image_caption)
27  plt.show()
```

Looking at the image in *Figure 6.3*, even for a human it is hard to tell whether there is a t-shirt, shirt, or bag in the image:

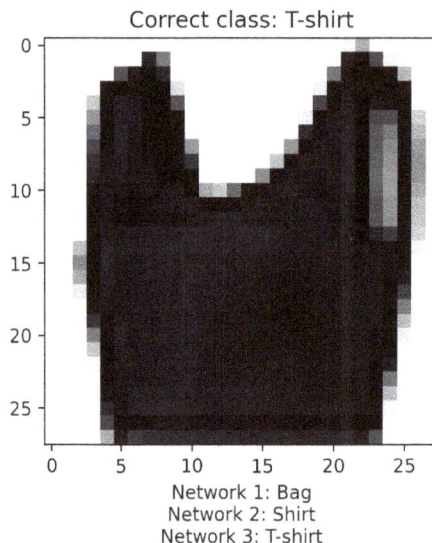

Figure 6.3: Image with highest variance among ensemble predictions. The correct ground truth label is "t-shirt," but it is hard to tell, even for a human

While we've seen that deep ensembles have several favorable qualities, they are not without limitations. In the next section, we'll explore what kinds of things we may wish to bear in mind when considering deep ensembles.

6.3.4   Practical limitations of deep ensembles

Some practical limitations of ensembles become obvious when taking them from the research environment to production at scale. We know that, in theory, the predictive performance and the uncertainty estimate of an ensemble is expected to improve as we add more ensemble members. However, there is a cost of adding more ensemble members as the memory footprint and inference cost of ensembles increases linearly with the number of ensemble members. This can make deploying ensembles in a production setting a costly choice. For every NN that we add to the ensemble, we will need to store an extra set of network weights, which significantly increases memory requirements. Equally, for every network, we will also need to run an additional forward pass during inference. Even though the inferences of different networks can be run in parallel, and the impact on inference time can therefore be mitigated, such an approach will still require more compute resources than single models. As more compute resources tend to translate to higher costs, the decision of using an ensemble versus a single model will need to trade off the benefits of better performance and uncertainty estimation with the increase in cost.

Recent research has tried to address or mitigate these practical limitations. In an approach called BatchEnsembles ([?]), for example, all ensemble members share one underlying weight matrix. The final weight matrix for each ensemble member is obtained by element-wise multiplication of this shared weight matrix with a rank-one matrix that is unique to each ensemble member. This reduces the number of parameters that need to be stored for each additional ensemble member and thus reduces memory footprint. The ensemble's computational cost is also reduced because the BatchEnsembles can exploit vectorization, and the output for all ensemble members can be computed in a single forward pass. In a different approach, called **multi-input/multi-output processing (MIMO**; [?]), a single network is encouraged to learn several independent sub-networks. During training, multiple inputs are passed along with multiple, correspondingly labeled outputs. The network will, for example, be presented with three images: of a dog, a cat and a chicken. Corresponding output labels are passed and the network will need to learn to predict "dog" on its first output node, "cat" on its second output node, and "chicken" on its third. During inference, one single image will be repeated three times and the MIMO ensemble will produce three different predictions (one on each output node). As a result, the memory footprint and computational cost of the MIMO approach is almost as little as that of a single neural network, while still providing all the benefits of an ensemble.

6.4 Exploring neural network augmentation with Bayesian last-layer methods

Through the course of *Chapter 5* and *Chapter 6*, we've explored a variety of methods for Bayesian inference with DNNs. These methods have incorporated some form of uncertainty information at every layer, whether through the use of explicitly probabilistic means or via ensemble-based or dropout-based approximations. These methods have certain advantages. Their consistent Bayesian (or, more accurately, approximately Bayesian) mechanics mean that they are consistent: the same principles are applied at each layer, both in terms of network architecture and update rules. This makes them easier to justify from a theoretical standpoint, as we know that any theoretical guarantees apply at each layer. In addition to this, it means that we have the benefit of being able to access uncertainties at every level: we can exploit embeddings in these networks just as we exploit embeddings in standard deep learning models, and we'll have access to uncertainties along with those embeddings.

However, these networks also come with some drawbacks. As we've seen, methods such as PBP and BBB have more complicated mechanics, which makes them more difficult to apply to more sophisticated neural network architectures. The topics earlier in this chapter demonstrate that we can get around this by using MC dropout or deep ensembles, but they increase our overheads in terms of computation and/or memory footprint. This is where **Bayesian Last-Layer (BLL)** methods (see *Figure 6.4*) come in. This class of methods gives us both the flexibility of using any NN architecture, while also being more computationally and memory efficient than MC dropout or deep ensembles.

Vanilla DNN DNN with last-layer
Bayesian regression

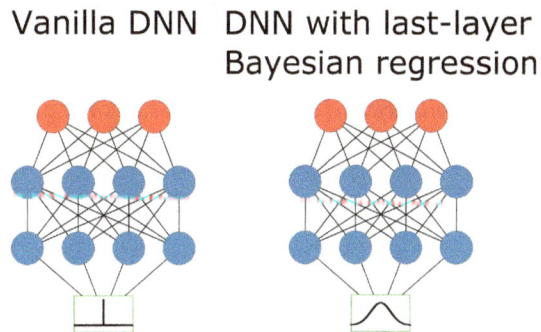

Figure 6.4: Vanilla NN compared to a BLL network

As you've probably guessed, the fundamental principle behind BLL methods is to estimate uncertainties only at the last-layer. But what you may not have guessed is why this is possible. Deep learning's success is due to the non-linear nature of NNs: the successive layers of non-linear transformations enable them to learn rich lower-dimensional representations of high-dimensional data. However, this non-linearity makes model uncertainty estimation difficult. Closed-form solutions for model uncertainty estimation are available for a variety of linear models, but unfortunately, this isn't the case for our highly non-linear DNNs. So, what can we do?

Well, fortunately for us, the representations learned by the DNNs can also serve as inputs to simpler linear models. In this way, we let the DNN do the heavy lifting: condensing our high-dimensional input space down to a task-specific low-dimensional representation. Because of this, the penultimate layer in the NN is far easier to deal with; after all, in most cases our output is simply some linear transformation of this layer. This means we can apply a linear model to this layer, which in turn means we can apply closed-form solutions for model uncertainty estimation.

We can make use of other last-layer approaches too; recent work has demonstrated that MC dropout is effective when applied only at the last layer. While this still requires multiple forward passes, these forward passes only need to be done for a single layer, making them much more computationally efficient, particularly for larger models.

6.4.1   Last-layer methods for Bayesian inference

The method proposed by Jasper Snoek et al. in their 2015 paper, *Scalable Bayesian Optimization Using Deep Neural Networks*, introduces the concept of using a post-hoc Bayesian linear regressor to obtain model uncertainties for DNNs. This method was devised as a way of achieving Gaussian Process-like high-quality uncertainties with improved scalability.

The method first involves training a NN on some data X and targets $\mathbf{y}$. This training phase trains a linear output layer, $\mathbf{z}_i$, resulting in a network that produces point estimates (typical of a standard DNN). We then take the penultimate layer (or the last hidden layer), $\mathbf{z}_{i-1}$, as our set of basis functions. From here, it's simply a case of replacing our final layer with a Bayesian linear regressor. Now, instead of our point estimates, our network will produce a predictive mean and variance. For further details on this method and adaptive basis regression, we point the reader to Jasper Snoek et al.'s paper, and to Christopher Bishop's *Pattern Recognition and Machine Learning*.

Now, let's see how we achieve this in code.

Step 1: Creating and training our base model

First, we set up and train our network:

```
1  from tensorflow.keras import Model, Sequential, layers, optimizers, metrics, losses
2  import tensorflow as tf
3  import tensorflow_probability as tfp
4  from sklearn.datasets import load_boston
5  from sklearn.model_selection import train_test_split
6  from sklearn.preprocessing import StandardScaler
7  from sklearn.metrics import mean_squared_error
8  import pandas as pd
9  import numpy as np
10
11  seed = 213
```

```
12  np.random.seed(seed)
13  tf.random.set_seed(seed)
14  dtype = tf.float32
15
16  boston = load_boston()
17  data = boston.data
18  targets = boston.target
19
20  X_train, X_test, y_train, y_test = train_test_split(data, targets, test_size=0.2)
21
22  # Scale our inputs
23  scaler = StandardScaler()
24  X_train = scaler.fit_transform(X_train)
25  X_test = scaler.transform(X_test)
26
27  model = Sequential()
28  model.add(layers.Dense(20, input_dim=13, activation='relu', name='layer_1'))
29  model.add(layers.Dense(8, activation='relu', name='layer_2'))
30  model.add(layers.Dense(1, activation='relu', name='layer_3'))
31
32  model.compile(optimizer=optimizers.Adam(),
33                loss=losses.MeanSquaredError(),
34                metrics=[metrics.RootMeanSquaredError()],)
35
36  num_epochs = 200
37  model.fit(X_train, y_train, epochs=num_epochs)
38  mse, rmse = model.evaluate(X_test, y_test)
```

Step 2: Using a neural network layer as a basis function

Now that we have our base network, we just need to access the penultimate layer so that we can feed this as our basis function to our Bayesian regressor. This is easily done using TensorFlow's high-level API, for example:

```
1  basis_func = Model(inputs=self.model.input,
2                        outputs=self.model.get_layer('layer_2').output)
```

This will build a model that will allow us to obtain the output of the second hidden layer by simply calling its `predict` method:

```
1  layer_2_output = basis_func.predict(X_test)
```

This is all that's needed to prepare our basis function for passing to our Bayesian linear regressor.

Step 3: Preparing our variables for Bayesian linear regression

For the Bayesian regressor, we assume that our outputs, $y_i \in \mathbf{y}$, are conditionally normally distributed according to a linear relationship with our inputs, $\mathbf{x}_i \in X$:

$$\mathbf{y}_i = \mathcal{N}(\alpha + \mathbf{x}_i^\mathsf{T}\beta, \sigma^2) \tag{6.3}$$

Here, α is our bias term, β are our model coefficients, and σ^2 is the variance associated with our predictions. We'll also make some prior assumptions about these parameters, namely:

$$\alpha \approx \mathcal{N}(0,1) \tag{6.4}$$

$$\beta \approx \mathcal{N}(0,1) \tag{6.5}$$

$$\sigma^2 \approx |\mathcal{N}(0,1)| \qquad\qquad (6.6)$$

Note that equation 6.6 denotes the half-normal of a Gaussian distribution. To wrap up the Bayesian regressor in such a way that it's easy (and practical) to integrate it with our Keras model, we'll create a `BayesianLastLayer` class. This class will use the TensorFlow Probability library to allow us to implement the probability distributions and sampling functions we'll need for our Bayesian regressor. Let's walk through the various components of our class:

```python
class BayesianLastLayer():

    def __init__(self,
                  model,
                  basis_layer,
                  n_samples=1e4,
                  n_burnin=5e3,
                  step_size=1e-4,
                  n_leapfrog=10,
                  adaptive=False):
        # Setting up our model
        self.model = model
        self.basis_layer = basis_layer
        self.initialize_basis_function()
        # HMC Settings
        # number of hmc samples
        self.n_samples = int(n_samples)
        # number of burn-in steps
        self.n_burnin = int(n_burnin)
        # HMC step size
        self.step_size = step_size
```

```
23          # HMC leapfrog steps
24          self.n_leapfrog = n_leapfrog
25          # whether to be adaptive or not
26          self.adaptive = adaptive
```

As we see here, our class requires at least two arguments at instantiation: `model`, which is our Keras model, and `basis_layer`, which is the layer output we wanted to feed to our Bayesian regressor. The following arguments are all parameters for the **Hamiltonian Monte-Carlo (HMC)** sampling for which we define some default values. These values may need to be changed depending on the input. For example, for a higher dimensional input (for instance, if you're using `layer_1`), you may want to further reduce the step size and increase both the number of burn-in steps and the overall number of samples.

Step 4: Connecting our basis function model

Next, we simply define a few functions for creating our basis function model and for obtaining its outputs:

```
1   def initialize_basis_function(self):
2       self.basis_func = Model(inputs=self.model.input,
3                           outputs=self.model.get_layer(self.basis_layer).output)
4
5   def get_basis(self, X):
6       return self.basis_func.predict(X)
```

Step 5: Creating a method to fit our Bayesian linear regression parameters

Now things get a little more complicated. We need to define the `fit()` method, which will use HMC sampling to find our model parameters α, β, and σ^2. We'll provide an overview of what the code is doing here, but for more (hands-on) information on sampling, we direct the reader to *Bayesian Analysis with Python* by Osvaldo Martin.

Firstly, we define a joint distribution using the priors described in equations 4.3-4.5. Thanks to TensorFlow Probability's `distributions` module, this is pretty straightforward:

```
1    def fit(self, X, y):
2        X = tf.convert_to_tensor(self.get_basis(X), dtype=dtype)
3        y = tf.convert_to_tensor(y, dtype=dtype)
4        y = tf.reshape(y, (-1, 1))
5        D = X.shape[1]
6
7        # Define our joint distribution
8        distribution = tfp.distributions.JointDistributionNamedAutoBatched(
9            dict(
10               sigma=tfp.distributions.HalfNormal(scale=tf.ones([1])),
11               alpha=tfp.distributions.Normal(
12                   loc=tf.zeros([1]),
13                   scale=tf.ones([1]),
14               ),
15               beta=tfp.distributions.Normal(
16                   loc=tf.zeros([D,1]),
17                   scale=tf.ones([D,1]),
18               ),
19               y=lambda beta, alpha, sigma:
20                   tfp.distributions.Normal(
21                       loc=tf.linalg.matmul(X, beta) + alpha,
22                       scale=sigma
23                   )
24           )
25       )
26    . . .
```

We then go on to set up our sampler using TensorFlow Probability's `HamiltonianMonteCarlo` sampler class. To do this, we'll need to define our target log probability function. The `distributions` module makes this fairly trivial, but we still need to define a function to feed our model parameters to the distribution object's `log_prob()` method (line 28). We can then pass this to the instantiation of `hmc_kernel`:

```
1   . . .

2           # Define the log probability function
3           def target_log_prob_fn(beta, alpha, sigma):
4               return distribution.log_prob(beta=beta, alpha=alpha, sigma=sigma, y=y)

5

6           # Define the HMC kernel we'll be using for sampling
7           hmc_kernel  = tfp.mcmc.HamiltonianMonteCarlo(
8             target_log_prob_fn=target_log_prob_fn,
9             step_size=self.step_size,
10            num_leapfrog_steps=self.n_leapfrog
11          )

12

13          # We can use adaptive HMC to automatically adjust the kernel step size
14          if self.adaptive:
15              adaptive_hmc = tfp.mcmc.SimpleStepSizeAdaptation(
16                  inner_kernel = hmc_kernel,
17                  num_adaptation_steps=int(self.n_burnin * 0.8)
18              )
19   . . .
```

Now that things are set up, we're ready to run our sampler. To do this, we call the `mcmc.sample_chain()` function, passing in our HMC parameters, an initial state for our model parameters, and our HMC sampler. We then run our sampling, which returns `states`, which comprises our parameter samples, and `kernel_results`, which contains some information about the sampling process. The information we care about here is to do with the proportion of accepted samples. If our sampler has run successfully, then we'll have a good proportion of accepted samples (indicating a high acceptance rate). If it hasn't been successful, then our acceptance rate will be low (perhaps even 0%!) and we may need to tune our sampler parameters. We print this to the console so that we can keep an eye on the acceptance rate (we wrap the call to `sample_chain()` in a `run_chain()` function so that it can be extended to sampling with multiple chains):

```python
. . .
        # If we define a function, we can extend this to multiple chains.
        @tf.function
        def run_chain():
            states, kernel_results = tfp.mcmc.sample_chain(
                num_results=self.n_samples,
                num_burnin_steps=self.n_burnin,
                current_state=[
                    tf.zeros((X.shape[1],1), name='init_model_coeffs'),
                    tf.zeros((1), name='init_bias'),
                    tf.ones((1), name='init_noise'),
                ],
                kernel=hmc_kernel
            )
            return states, kernel_results

        print(f'Running HMC with {self.n_samples} samples.')
        states, kernel_results = run_chain()

        print('Completed HMC sampling.')
        coeffs, bias, noise_std = states
        accepted_samples = kernel_results.is_accepted[self.n_burnin:]
        acceptance_rate = 100*np.mean(accepted_samples)
        # Print the acceptance rate - if this is low, we need to check our
        # HMC parameters
        print('Acceptance rate: %0.1f%%' % (acceptance_rate))
```

Once we've run our sampler, we can fetch our model parameters. We take them from the post-burn-in samples and assign them to class variables for later use in inference:

```
1          # Obtain the post-burnin samples
2          self.model_coeffs = coeffs[self.n_burnin:,:,0]
3          self.bias = bias[self.n_burnin:]
4          self.noise_std = noise_std[self.n_burnin:]
```

Step 6: Inference

The last thing we need to do is implement a function to make predictions using the
learned parameters of our joint distribution. To do this, we'll define two functions:
get_pred_dist(), which will obtain the posterior predictive distribution given our
input, and predict(), which will call get_pred_dist() and compute our mean (μ) and
standard deviation (σ) from our posterior distribution:

```
1     def get_pred_dist(self, X):
2         predictions = (tf.matmul(X, tf.transpose(self.model_coeffs)) +
3                       self.bias[:,0])
4         noise = (self.noise_std[:,0] *
5                  tf.random.normal([self.noise_std.shape[0]]))
6         return predictions + noise
7
8     def predict(self, X):
9         X = tf.convert_to_tensor(self.get_basis(X), dtype=dtype)
10        pred_dist = np.zeros((X.shape[0], self.model_coeffs.shape[0]))
11        X = tf.reshape(X, (-1, 1, X.shape[1]))
12        for i in range(X.shape[0]):
13            pred_dist[i,:] = self.get_pred_dist(X[i,:])
14
15        y_pred = np.mean(pred_dist, axis=1)
16        y_std = np.std(pred_dist, axis=1)
17        return y_pred, y_std
```

And that's it! We have our BLL implementation! With this class, we have a powerful and principled means of obtaining Bayesian uncertainty estimates by using penultimate NN layers as basis functions for Bayesian regression. Making use of it is as simple as passing our model and defining which layer we want to use as our basis function:

```
1  bll = BayesianLastLayer(model, 'layer_2')
2
3  bll.fit(X_train, y_train)
4
5  y_pred, y_std = bll.predict(X_test)
```

While this is a powerful tool, it's not always suited for the task at hand. You can experiment with this yourself: try creating a model with a larger embedding layer. As the size of the layer increases, you should start to see that the acceptance rate of the sampler drops. Once it's large enough, the acceptance rate may even fall to 0%. So, we'll need to modify the parameters of our sampler: reducing the step size, increasing the number of samples, and increasing the number of burn-in samples. As the dimensionality of the embedding grows, it becomes more and more difficult to obtain a representative set of samples for the distribution.

For some applications, this isn't an issue, but when dealing with complex, high-dimensional data, this can quickly become problematic. Applications in domains such as computer vision, speech processing, and molecular modeling all rely on high-dimensional embeddings. One solution here is to reduce these embeddings further, for example, via dimensionality reduction. But doing so can have an unpredictable effect on these encodings: in fact, by reducing the dimensionality, you could be unintentionally removing sources of uncertainty, resulting in poorer quality uncertainty estimates.

So, what can we do instead? Fortunately, there are a few other last-layer options we can employ. Next, we'll see how we can use last-layer dropout to approximate the Bayesian linear regression approach introduced here.

6.4.2   Last-layer MC dropout

Earlier in the chapter, we saw how we can use dropout at test time to obtain a distribution over our model predictions. Here, we'll combine that concept with the concept of last-layer uncertainties: adding an MC dropout layer, but only as a single layer that we add to a pre-trained network.

Step 1: Connecting to our base model

Similarly to the Bayesian last-layer method, we first need to obtain the output from our model's penultimate layer:

```
1  basis_func = Model(inputs=model.input,
2                      outputs=model.get_layer('layer_2').output)
```

Step 2: Adding an MC dropout layer

Now, instead of implementing a Bayesian regressor, we'll simply instantiate a new output layer, which applies dropout to the penultimate layer:

```
1  ll_dropout = Sequential()
2  ll_dropout.add(layers.Dropout(0.25))
3  ll_dropout.add(layers.Dense(1, input_dim=8, activation='relu', name='dropout_layer')
```

Step 3: Training the MC dropout last-layer

Because we've now added a new final layer, we need to run an additional step of training so that it can learn the mapping from our penultimate layer to the new output; but because our original model is doing all of the heavy lifting, this training is both computationally cheap and quick to run:

```
1  ll_dropout.compile(optimizer=optimizers.Adam(),
2                      loss=losses.MeanSquaredError(),
3                      metrics=[metrics.RootMeanSquaredError()],)
4  num_epochs = 50
5  ll_dropout.fit(basis_func.predict(X_train), y_train, epochs=num_epochs)
```

Step 4: Obtaining uncertainties

Now that our last layer is trained, we can implement a function to obtain the mean and standard deviation for our predictions using multiple forward passes of our MC dropout layer; line 3 onwards should be familiar from earlier in the chapter, and line 2 simply obtains the output from our original model's penultimate layer:

```
1  def predict_ll_dropout(X, basis_func, ll_dropout, nb_inference):
2      basis_feats = basis_func(X)
3      ll_pred = [ll_dropout(basis_feats, training=True) for _ in range(nb_inference)]
4      ll_pred = np.stack(ll_pred)
5      return ll_pred.mean(axis=0), ll_pred.std(axis=0)
```

Step 5: Inference

All that's left is to call this function and obtain our new model outputs, complete with uncertainty estimates:

```
1  y_pred, y_std = predict_ll_dropout(X_test, basis_func, ll_dropout, 50)
```

Last-layer MC dropout is by far one of the easiest ways to obtain uncertainty estimates from pre-trained networks. Unlike standard MC dropout, it doesn't require training a model from scratch, so you can apply this post-hoc to networks you've already trained. Additionally, unlike the other last-layer methods, it can be implemented in just a few straightforward steps that never stray from TensorFlow's standard API.

6.4.3 Recap of last-layer methods

Last-layer methods are an excellent tool for when you need to obtain uncertainty estimates from a pre-trained network. Given how expensive and time-consuming neural network training can be, it's nice not to have to start from scratch just because you need some predictive uncertainties. Additionally, given that more and more machine learning practitioners are relying on state-of-the-art pre-trained models, these kinds of techniques are a practical way to incorporate model uncertainties after the fact.

But there are drawbacks to last-layer methods too. Unlike other methods, we're relying on a fairly limited source of variance: the penultimate layer of our model. This limits how much stochasticity we can induce over our model outputs, meaning we're at risk of over-confident predictions. Bear this in mind when using last-layer methods and, if you see the hallmark signs of over-confidence, consider using a more comprehensive method to obtain your predictive uncertainties.

6.5 Summary

In this chapter, we've seen how familiar machine learning and deep learning concepts can be used to develop models with predictive uncertainties. We've also seen how, with relatively minor modifications, we can add uncertain estimates to pre-trained models. This means we can go beyond the point-estimate approach of standard NNs, using uncertainties to gain valuable insights into the performance of our models, and allowing us to develop more robust applications.

However, as with the methods introduced in *Chapter 5*, all techniques have advantages and disadvantages. For example, last-layer methods may give us the flexibility to add uncertainties to any model, but they're limited by the representation that the model has already learned. This could result in very low variance outputs, resulting in an overconfident model. Similarly, while ensemble methods allow us to capture variance across every layer of the network, they come at significant computational cost, requiring that we have multiple networks, rather than just a single network.

In the next chapter, we will examine the advantages and disadvantages in more detail, and learn how we can address some of the shortcomings of these methods.

7

Practical Considerations for Bayesian Deep Learning

Over the last two chapters, *Chapter 5* and *Chapter 6*, we've been introduced to a range of methods that facilitate Bayesian inference with neural networks. *Chapter 5* introduced specially crafted Bayesian neural network approximations, while *Chapter 6* showed how we can use the standard toolbox of machine learning to add uncertainty estimates to our models. These families of methods come with their own advantages and disadvantages. In this chapter, we will explore some of these differences in practical scenarios in order to help you understand how to select the best method for the task at hand.

We will also look at different sources of uncertainty, which can improve your understanding of the data or help you choose a different exception path based on the source of uncertainty. For example, if a model is uncertain because the input data is inherently noisy, you might want to send the data to a human for review. However, if a model is uncertain because the input data has not been seen before, it might be helpful to add this data to your model so that it can reduce its uncertainty on this type of data.

Bayesian deep learning techniques can help you to distinguish between these sources of uncertainty. These topics will be covered in the following sections:

- Balancing uncertainty quality and computational considerations

- BDL and sources of uncertainty

7.1   Technical requirements

To complete the practical tasks in this chapter, you will need a Python 3.8 environment with the SciPy and scikit-learn stack and the following additional Python packages installed:

- TensorFlow 2.0

- TensorFlow Probability

All of the code for this book can be found on the GitHub repository for the book: `https://github.com/PacktPublishing/Enhancing-Deep-Learning-with-Bayesian-Inference`.

7.2   Balancing uncertainty quality and computational considerations

While Bayesian methods have many benefits, there are also trade-offs to consider in terms of memory and computational overheads. These considerations play a critical role in selecting the most appropriate methods to use within real-world applications.

In this section, we'll examine the trade-offs between different methods in terms of performance and uncertainty quality, and we'll learn how we can use TensorFlow's profiling tools to measure the computational costs associated with different models.

7.2.1   Setting up our experiments

To evaluate the performance of different models, we'll need a few different datasets. One of these is the California Housing dataset, which is conveniently provided by scikit-learn. The others we'll use are commonly used in papers comparing uncertainty models: the Wine Quality dataset and the Concrete Compressive Strength dataset. Let's take a look at a breakdown of these datasets:

- **California Housing**: This dataset comprises a number of features for different regions in California derived from the 1990 California census. The dependent variable is house value, which is provided as the median value for each block of houses. In older papers, you'll see the Boston Housing dataset used; the California Housing dataset is now favored due to ethical issues around the Boston Housing dataset.

- **Wine Quality**: The Wine Quality dataset comprises features pertaining to the chemical composition of a variety of different wines. The value we're trying to predict is the subjective quality of the wine.

- **Concrete Compressive Strength**: The Concrete Compressive Strength dataset's features describe the ingredients used for mixing concrete, and each data point is a different concrete mixture. The dependent variable is the concrete's compressive strength.

The following experiments will use code from the book's GitHub repository (https://github.com/PacktPublishing/Bayesian-Deep-Learning), which we've seen in various forms in the previous chapters. The example assumes that we're running the code from within this repository.

Importing our dependencies

As usual, we'll start by importing our dependencies:

```
1  import tensorflow as tf
2  import numpy as np
3  import matplotlib.pyplot as plt
4  import tensorflow_probability as tfp
5  from sklearn.metrics import accuracy_score, mean_squared_error
```

```
 6  from sklearn.datasets import fetch_california_housing, load_diabetes
 7  from sklearn.model_selection import train_test_split
 8  import seaborn as sns
 9  import pandas as pd
10  import os
11
12  from bayes_by_backprop import BBBRegressor
13  from pbp import PBP
14  from mc_dropout import MCDropout
15  from ensemble import Ensemble
16  from bdl_ablation_data import load_wine_quality, load_concrete
17  from bdl_metrics import likelihood
```

Here, we can see that we're using a number of model classes defined in the repository. While these classes each support different architectures of models, they'll be using the default structure, which is defined in constants.py. This structure comprises a single densely connected hidden layer of 64 units, and a single densely connected output layer. The BBB and PBP equivalents will be used and are defined as their default architectures in their respective classes.

Preparing our data and models

Now we need to prepare our data and models to run our experiments. Firstly, we'll set up a dictionary that we can iterate over to access data from different datasets:

```
1  datasets = {
2      "california_housing": fetch_california_housing(return_X_y=True, as_frame=True),
3      "diabetes": load_diabetes(return_X_y=True, as_frame=True),
4      "wine_quality": load_wine_quality(),
5      "concrete": load_concrete(),
6  }
```

Next, we'll create another dictionary to allow us to iterate over our different BDL models:

```
1  models = {
2      "BBB": BBBRegressor,
3      "PBP": PBP,
4      "MCDropout": MCDropout,
5      "Ensemble": Ensemble,
6  }
```

Finally, we'll create a dictionary to hold our results:

```
1  results = {
2      "LL": [],
3      "MSE": [],
4      "Method": [],
5      "Dataset": [],
6  }
```

Here, we see that we'll be recording two results: the log-likelihood, and the mean-squared error. We're using these metrics as we're looking at regression problems, but for classification problems you may opt to use F-score or accuracy in place of the mean-squared error, and expected calibration error in place of (or as well as) log-likelihood. We'll also be storing the model type in the Method field, and the dataset in the Dataset field.

Running our experiments

Now we're ready to run our experiments. However, we're not only interested in the model performance, but also the computational considerations of our various models. As such, we'll see calls to tf.profiler in the following code. First, however, we'll set a few parameters:

```
1  # Parameters
2  epochs = 10
3  batch_size = 16
4  logdir_base = "profiling"
```

Here, we're setting the number of epochs each model will train for, as well as the batch size each model will use. We're also setting `logdir_base`, the location that all of our profiling logs will be written to.

Now we're ready to drop in our experiment code. We'll start by iterating over the datasets:

```
1  for dataset_key in datasets.keys():
2      X, y = datasets[dataset_key]
3      X_train, X_test, y_train, y_test = train_test_split(X, y, test_size=0.33)
4      ...
```

Here, we see that for each dataset we're splitting the data, using $\frac{2}{3}$ of the data for training and $\frac{1}{3}$ for testing.

Next, we iterate over the models:

```
1      ...
2      for model_key in models.keys():
3          logdir = os.path.join(logdir_base, model_key + "_train")
4          os.makedirs(logdir, exist_ok=True)
5          tf.profiler.experimental.start(logdir)
6          ...
```

For each model, we instantiate a new log directory to log the training information. We
then instantiate the model and run `model.fit()`:

```
1        . . .
2        model = models[model_key]()
3        model.fit(X_train, y_train, batch_size=batch_size, n_epochs=epochs)
```

Once the model fits, we stop the profiler and create a new directory to log the prediction
information, after which we start the profiler again:

```
1        . . .
2        tf.profiler.experimental.stop()
3        logdir = os.path.join(logdir_base, model_key + "_predict")
4        os.makedirs(logdir, exist_ok=True)
5        tf.profiler.experimental.start(logdir)
6        . . .
```

With the profiler running, we run predict, after which we again stop the profiler. With
our predictions in hand, we can compute our mean squared error and log-likelihood, and
store these to our `results` dictionary. Finally, we run
`tf.keras.backend.clear_session()` to clear our TensorFlow graph after each
experiment within our `model` loop:

```
1        . . .
2        y_pred, y_var = model.predict(X_test)
3
4        tf.profiler.experimental.stop()
5
6        y_pred = y_pred.reshape(-1)
7        y_var = y_var.reshape(-1)
8
9        mse = mean_squared_error(y_test, y_pred)
```

```
10          ll = likelihood(y_test, y_pred, y_var)
11          results["MSE"].append(mse)
12          results["LL"].append(ll)
13          results["Method"].append(model_key)
14          results["Dataset"].append(dataset_key)
15          tf.keras.backend.clear_session()
16  ...
```

Once we've got results for all models and all datasets, we convert our results dictionary
into a pandas DataFrame:

```
1  ...
2  results = pd.DataFrame(results)
```

Now we're ready to analyze our data!

7.2.2 Analyzing model performance

With the data obtained from our experiments, we can plot this to see which models
performed best on which datasets. To do so, we'll use the following plotting code:

```
1  results['NLL'] = -1*results['LL']
2
3  i = 1
4  for dataset in datasets.keys():
5      for metric in ["NLL", "MSE"]:
6          df_plot = results[(results['Dataset']==dataset)]
7          df_plot = groupedvalues = df_plot.groupby('Method').sum().reset_index()
8          plt.subplot(3,2,i)
9          ax = sns.barplot(data=df_plot, x="Method", y=metric)
10         for index, row in groupedvalues.iterrows():
11             if metric == "NLL":
12                 ax.text(row.name, 0, round(row.NLL, 2),
```

```
13                                    color='white', ha='center')
14                    else:
15                        ax.text(row.name, 0, round(row.MSE, 2),
16                                    color='white', ha='center')
17            plt.title(dataset)
18            if metric == "NLL" and dataset == "california_housing":
19                plt.ylim(0, 100)
20            i+=1
21  fig = plt.gcf()
22  fig.set_size_inches(10, 8)
23  plt.tight_layout()
```

Note that initially we add a 'NLL' field to our pandas DataFrame. This gives us the negative log-likelihood. This makes things a little less confusing when looking at the plots, as lower is better for both mean squared error and negative log-likelihood.

The code iterates over the datasets and metrics, and creates some nice bar plots with the help of the Seaborn plotting library. In addition to this, we use calls to ax.text() to overlay the metric values on the bar plots, allowing us to see the values clearly.

Also notice how, for the California Housing data, we're capping our y values at 100 for our negative log-likelihood. This is because, for this dataset, our negative log-likelihood value is *incredibly* high - making it difficult to view this in context with the other values. This is another reason why we're overlaying the metric values, to allow us to compare them easily as one of the values exceeds the limit of the plot.

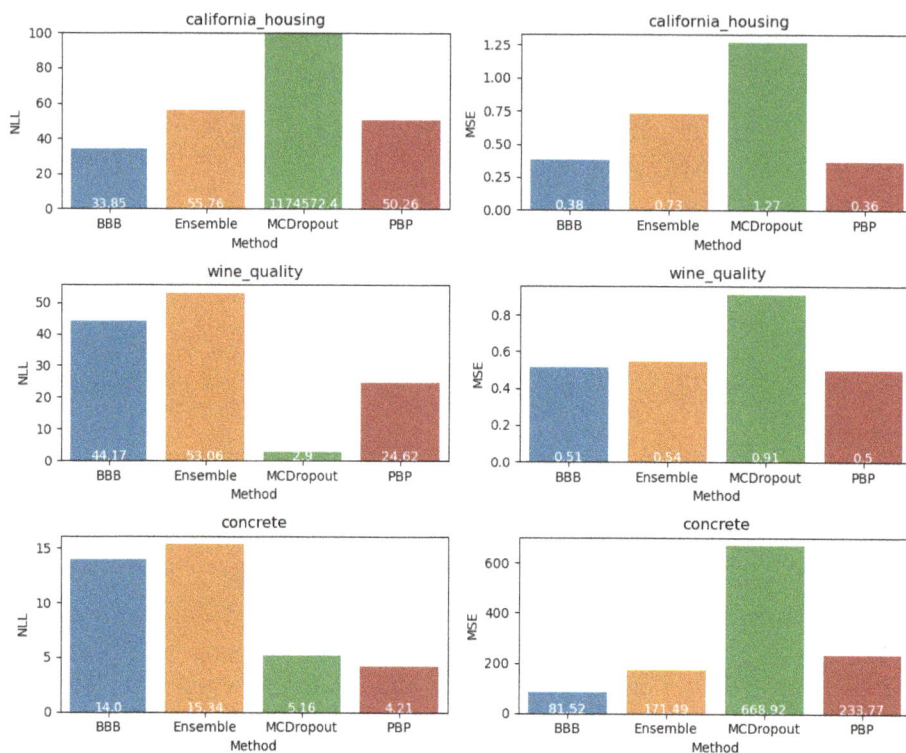

Figure 7.1: Bar plot of results from LL and MSE experiments

It's worth noting that, for fair comparison, we've used the equivalent architecture across all models, used the same batch size, and trained for the same number of epochs.

As we see here, there's no single best method: each model performs differently depending on the data, and a model with low mean squared error isn't guaranteed to also have a low negative log-likelihood score. Generally speaking, MC dropout exhibits the worst mean squared error scores; however, it also produces the best negative log-likelihood observed during our experiments for the Wine Quality dataset, for which it achieves a negative log-likelihood of 2.9. This is due to the fact that, while it generally performs worse in terms of error, its uncertainties are very high. As such, because it is more uncertain in regions for which it is wrong, it produces a more favorable negative log-likelihood score. We can see this if we plot the errors against the uncertainties:

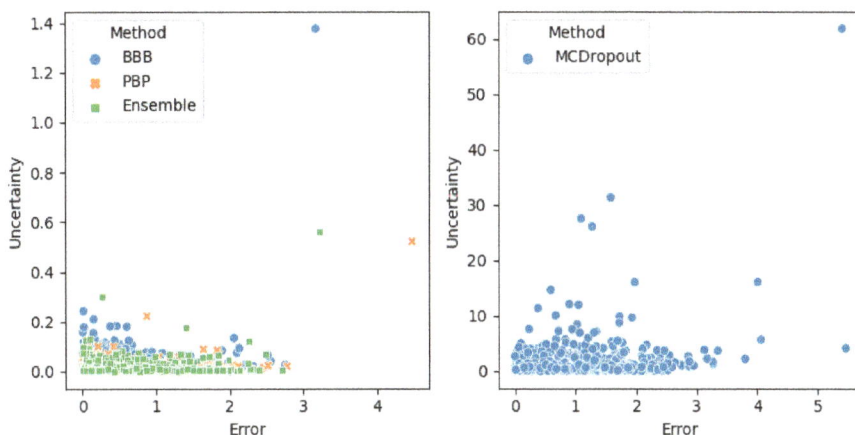

Figure 7.2: Scatter plot of errors vs uncertainty estimates

In *Figure 7.2*, we see the BBB, PBP, and ensemble results in the plot on the left, while MC dropout's results are in the plot on the right. The reason for this is that MC dropout's uncertainty estimates are two orders of magnitude higher than the uncertainty estimates produced by the other methods, thus they can't be clearly represented on the same axes. These very high-magnitude uncertainties are also the reason behind its comparatively low negative log-likelihood score. This is quite a surprising example for MC dropout, as it is typically *over-confident*, whereas in this case, it's clearly *under-confident*.

While MC dropout's under-confidence may lead to better likelihood scores, these metrics need to be considered in context; we typically want a good trade-off between likelihood and error. As such, PBP is probably the best choice in the case of the Wine Quality data as it has the lowest error, but it also has a reasonable likelihood; its negative log-likelihood is not so low as to be suspicious, but also low enough to know that the uncertainty estimates will be reasonably consistent and principled.

In the case of the other datasets, the choices are a little more straightforward: BBB is the clear winner for California Housing, and PBP again proves to be the sensible choice on balance in the case of Concrete Compressive Strength. This is all with the important caveat that none of these networks have been specifically optimized for these datasets:

this is merely an illustrative example.

Crucially, it's going to come down to the specific application and how much robust uncertainty estimates matter. For example, in a safety-critical scenario, you'll want to go with a method with the most robust uncertainty estimates, and so you may favor under-confidence over a lower error because you want to make sure that you're only going with the model when you're very confident in its outcome. In these cases, you may well go for an under-confident but high likelihood (low negative likelihood) method such as MC dropout on the Wine Quality dataset.

In other cases, perhaps uncertainty doesn't matter at all, in which case you may just go for a standard neural network. But in most mission-critical or safety-critical applications, you're going to want to strike a balance and take advantage of the additional information provided by model uncertainty estimates while still achieving a low error score. However, practically, these performance metrics aren't the only thing we need to consider when developing machine learning systems. We also care about the practical implications. In the next section, we'll see how the computational requirements of these models stack up against each other.

7.2.3 Computational considerations of Bayesian deep learning models

For every real-world application of machine learning, there are considerations beyond performance: we also need to understand the practical limitations of the compute infrastructure. They are usually governed by a few things, but existing infrastructure and cost tend to come up time and time again.

Existing infrastructure is often important because unless it's a totally new project, it's a case of working out how a machine learning model can be integrated, and this means either finding or requesting additional computational resources on a hardware or software stack. It will come as no surprise that cost is a significant factor: every project has a budget, and the expense that the machine learning component of the solution brings to the table needs to be balanced against the advantages that it provides. The budget will often dictate what machine learning solutions are feasible based on the cost of the computational resources required to train, deploy, and run them at inference time.

To get some insight into how the methods here compare in terms of their computational requirements, we'll look at the output of the TensorFlow profiler that we included in our experiment code. To do this, we simply need to run TensorBoard from the command line, pointing to the logging directory for the particular model we're interested in:

```
1  tensorboard --logdir profiling/BBB_train/
```

This will start a TensorBoard instance (typically at `http://localhost:6006/`). Copy the URL into your browser, and you'll be greeted with the TensorBoard GUI. TensorBoard provides you with a suite of tools for understanding the performance of your TensorFlow models, from the execution time right down to the memory allocation of different processes. You can scroll through the available tools via the **Tools** selection box in the top left corner of the screen:

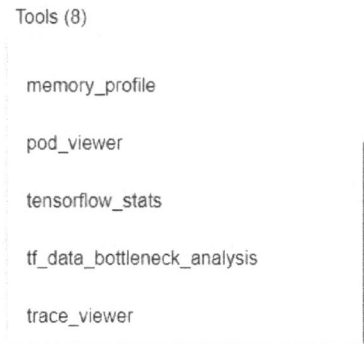

Tools (8)

memory_profile

pod_viewer

tensorflow_stats

tf_data_bottleneck_analysis

trace_viewer

Figure 7.3: Tool selection box in the TensorBoard GUI

To get a very detailed picture of what's going on, take a look at the Trace Viewer:

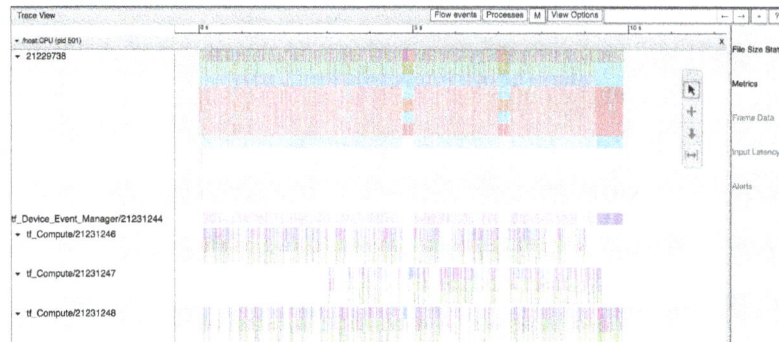

Figure 7.4: Trace Viewer in the TensorBoard GUI

Here, we get an overall picture of the time taken to run our model's functions, as well as a detailed picture of which processes are running under the hood, and how long each of these processes take to run. We can even dig deeper by double-clicking on a block and looking at its statistics. For example, we can double-click on the **train** block:

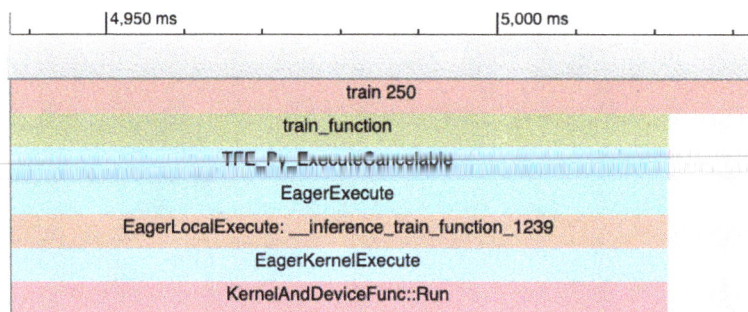

Figure 7.5: Trace Viewer from the TensorBoard GUI, highlighting the train block

This brings up some information at the bottom of our screen. This allows us to closely examine the run time of this process. If we click on **Duration**, we get a detailed breakdown of the run duration statistics for the process:

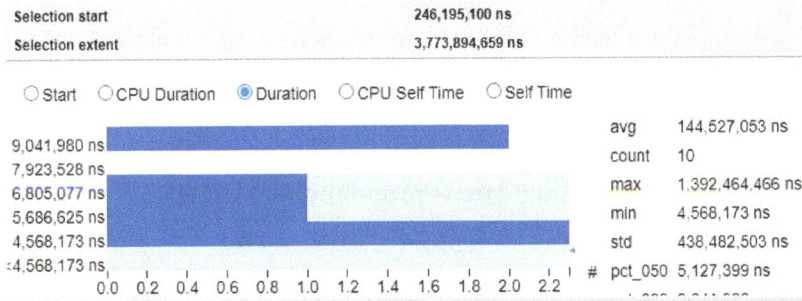

Figure 7.6: Examining the statistics of a block in the TensorBoard Trace Viewer

Here, we see that the process was run 10 times (once per epoch) and that the average
duration is 144,527,053 ns (nanoseconds). Let's use our profiler results for the Concrete
Compression Strength dataset and collect the runtime and memory allocation
information using TensorBoard. If we do this for each of our models' training runs, we
obtain the following information:

Profiling data for model training		
Model	**Peak memory usage (MiB)**	**Duration (ms)**
BBB	0.09	4270
PBP	0.253	10754
MCDropout	0.126	2198
Ensemble	0.215	20630

*Figure 7.7: Table of profiling data for model training for Concrete Compressive Strength
dataset*

Here, we see that MC dropout is the fastest model to train for this dataset, taking half as
long as BBB. We also see that the ensemble takes by far the longest to train, nearly 10
times as long as MC dropout, despite the fact that the ensemble comprises only 5 models.
In terms of memory usage, we see that the ensemble again performs poorly, but that
PBP is the most memory hungry of the models, while BBB has the lowest peak memory
usage.

But it's not just training that counts. We also need to factor in the computational cost of inference. Looking at our profiling data for our models' predict functions, we see the following:

Profiling data for model predictions

Model	Peak memory usage (MiB)	Duration (ms)
BBB	0.116	849
PBP	1.27	176
MCDropout	0.548	23
Ensemble	0.389	17

Figure 7.8: Table of profiling data for model prediction for Concrete Compressive Strength dataset

Interestingly, here we see that the ensemble is in the lead when it comes to model inference speed, and it also comes in second when we look at peak memory usage for predictions. In contrast, PBP has by far the highest peak memory usage, while BBB takes the longest to run inference.

There are various factors contributing to the results we see here. Firstly, it's important to note that none of these models are properly optimized for computational performance. For example, we could significantly cut down the training duration for our ensemble by training all the ensemble members in parallel, which we don't do here. Similarly, because PBP used a lot of high-level code in its implementation (unlike the other methods, which are all built on nicely optimized TensorFlow or TensorFlow Probability code), its performance suffers as a result.

Most crucially, we need to ensure that, when selecting the right model for the job, we consider the computational implications as well as the typical performance metrics. So, with all that in mind, how do we choose the right model?

7.2.4 Choosing the right model

With our performance metrics and profiling information in hand, we have all the data we need to choose the right model for the task. But model selection isn't easy; we see here that all our models have advantages and disadvantages.

If we start with performance metrics, then we see that BBB has the lowest mean squared error, but it also has a very high negative log-likelihood. So, the best choice on the basis of performance metrics alone is PBP: it has the lowest negative log-likelihood score, and its mean squared error is nowhere near as poor as MC dropout's error, making PBP the best choice, on balance.

However, if we look at the computational implications in *Figure 7.7* and *7.8*, we see that PBP is one of the worst choices both in terms of memory usage and execution time. The best choice here, on balance, would be MC dropout: its prediction time is only a little slower than the prediction time of the ensemble, and it has the shortest training duration.

At the end of the day, it's entirely down to the application: perhaps inference doesn't need to be run in real time, so we can go with our PBP implementation. Or perhaps inference time and low error are our key considerations, in which case the ensemble is a good choice. As we see here, metrics and computational overheads need to be considered in context, and, as with any class of machine learning models, there's no single best choice for all applications. It's all down to choosing the right tool for the job.

In this section, we've introduced tools for comprehensively understanding model performance and demonstrated how important it is to consider a range of factors when selecting a model. Fundamentally, performance analysis and profiling are as important for helping us to make the right practical choices as they are for helping us to uncover opportunities for further improvements. We may not have time for further optimizing our code and so may need to be pragmatic and go with the best computationally optimized method we have to hand. Alternatively, the business case may dictate that we need the model with the best performance, which may justify investing time to optimize our code and reduce the computational overheads of a given method. In the next section, we'll take a look at another important practical consideration of working with BDL methods as we learn how we can use these methods to better understand sources of uncertainty.

7.3   BDL and sources of uncertainty

In this case study, we will look at how we can model aleatoric and epistemic uncertainty in a regression problem when we are trying to predict a continuous outcome variable. We will use a real-life dataset of diamonds that contains the physical attributes of more than 50,000 diamonds as well as their prices. In particular, we will look at the relationship between the weight of a diamond (measured as its **carat**) and the price paid for the diamond.

Step 1: Setting up the environment

To set up the environment, we import several packages. We import `tensorflow` and `tensorflow_probability` for building and training vanilla and probabilistic neural networks, `tensorflow_datasets` for importing the diamonds data set, `numpy` for performing calculations and operations on numerical arrays (such as calculating the mean), `pandas` for handling DataFrames, and `matplotlib` for plotting:

```
1  import matplotlib.pyplot as plt
2  import numpy as np
3  import pandas as pd
4  import tensorflow as tf
5  import tensorflow_probability as tfp
6  import tensorflow_datasets as tfds
```

First, we load the diamonds dataset using the `load` function provided by `tensorflow_datasets`. We load the dataset as a `pandas` DataFrame, which is convenient for preparing the data for training and inference.

```
1  ds = tfds.load('diamonds', split='train')
2  df = tfds.as_dataframe(ds)
```

The dataset contains many different attributes of diamonds, but here we will focus on carat and price by selecting the respective columns from the DataFrame:

```
1  df = df[["features/carat", "price"]]
```

We then divide the dataset into train and test splits. We use 80% of the data for training and 20% for testing:

```
1  train_df = df.sample(frac=0.8, random_state=0)
2  test_df = df.drop(train_df.index)
```

For further processing, we convert the train and test DataFrames to NumPy arrays:

```
1  carat = np.array(train_df['features/carat'])
2  price = np.array(train_df['price'])
3  carat_test = np.array(test_df['features/carat'])
4  price_test = np.array(test_df['price'])
```

We also save the number of training samples into a variable because we will need it later during model training:

```
1  NUM_TRAIN_SAMPLES = carat.shape[0]
```

Finally, we define a plotting function. This function will come in handy during the rest of the case study. It allows us to plot the data points as well as the fitted model predictions and their standard deviations:

```
1   def plot_scatter(x_data, y_data, x_hat=None, y_hats=None, plot_std=False):
2       # Plot the data as scatter points
3       plt.scatter(x_data, y_data, color="k", label="Data")
4       # Plot x and y values predicted by the model, if provided
5       if x_hat is not None and y_hats is not None:
6           if not isinstance(y_hats, list):
7               y_hats = [y_hats]
8           for ind, y_hat in enumerate(y_hats):
9               plt.plot(
10                  x_hat,
11                  y_hat.mean(),
12                  color="#e41a1c",
13                  label="Prediction" if ind == 0 else None,
14              )
15      # Plot standard deviation, if requested
16      if plot_std:
17          for ind, y_hat in enumerate(y_hats):
18              plt.plot(
19                  x_hat,
20                  y_hat.mean() + 2 * y_hat.stddev(),
21                  color="#e41a1c",
22                  linestyle="dashed",
23                  label="Prediction + stddev" if ind == 0 else None,
24              )
25              plt.plot(
26                  x_hat,
27                  y_hat.mean() - 2 * y_hat.stddev(),
28                  color="#e41a1c",
29                  linestyle="dashed",
30                  label="Prediction - stddev" if ind == 0 else None,
31              )
32      # Plot x- and y-axis labels as well as a legend
```

```
33      plt.xlabel("carat")
34      plt.ylabel("price")
35      plt.legend()
```

Using this function, we can have a first look at the training data by running the following:

```
1   plot_scatter(carat, price)
```

The training data distribution is shown in *Figure 7.9*. We observe that the relationship between carat and diamond price is non-linear, with prices increasing more rapidly at higher carat.

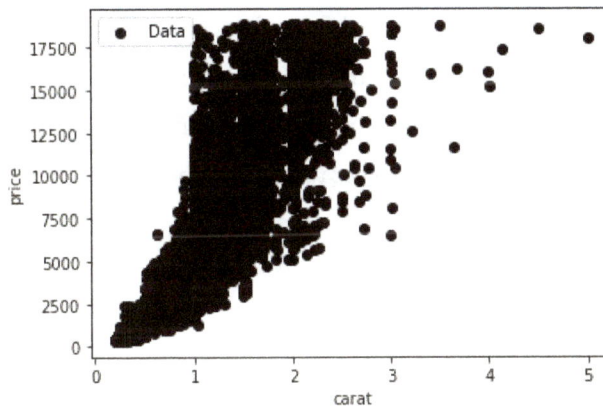

Figure 7.9: Relationship between the carat of a diamond and its price

Step 2: Fitting a model without uncertainty

Having completed the setup, we are ready to fit some regression models to the data. We start by fitting a neural network model without quantifying the uncertainty in the predictions. This allows us to establish a baseline and to introduce some tools (in the form of functions) that will be useful for all of the models in this case study.

It is recommended that you normalize the input features to a neural network model. In this example, that means normalizing the weight of the diamonds in carats. Normalizing input features will make the model converge faster during training. `tensorflow.keras` provides a convenient normalization function that allows us to do just that. We can use it as follows:

```
1  normalizer = tf.keras.layers.Normalization(input_shape=(1,), axis=None)
2  normalizer.adapt(carat)
```

We will also need a loss function, ideally one that can be used for all the models in this case study. A regression model can be posed as $P(y|x, w)$, the probability distribution of labels y given the inputs x and model parameters w. We can fit such a model to the data by minimizing the negative log-likelihood loss $-logP(y|x)$. In Python code, this can be written as a function that takes as input the true outcome value y_true and the predicted outcome distribution y_pred and returns the negative log-likelihood of the outcome value under the predicted outcome distribution, which is implemented in the `log_prob()` method provided by the `distributions` module in `tensorflow_probability`:

```
1  def negloglik(y_true, y_pred):
2      return -y_pred.log_prob(y_true)
```

Equipped with these tools, let's build our first model. We use the normalizer function that we just defined to normalize the model inputs. We then stack two dense layers on top. The first dense layer consists of 32 nodes. This allows us to model the non-linearity observed in the data. The second dense layer consists of one node in order to reduce the model prediction to a single value. Importantly, we do not use the output produced by this second dense layer as the model output. Instead, we use the dense layer output to parameterize the mean of a normal distribution, which means that we are modeling the ground truth labels using a normal distribution. We also set the variance of the normal distribution to 1. Parameterizing the mean of a distribution while setting the variance to a fixed value implies that we are modeling the overall trend of the data without yet quantifying uncertainty in the model's predictions:

```
1   model = tf.keras.Sequential(
2       [
3           normalizer,
4           tf.keras.layers.Dense(32, activation="relu"),
5           tf.keras.layers.Dense(1),
6           tfp.layers.DistributionLambda(
7               lambda t: tfp.distributions.Normal(loc=t, scale=1)
8           ),
9       ]
10  )
```

As we've seen in previous case studies, to train the model we use the `compile()` and `fit()` functions. During compilation of the model, we specify the `Adam` optimizer and the previously defined loss function. For the `fit` function, we specify that we want to train the model for 100 epochs on the carat and price data:

```
1   # Compile
2   model.compile(optimizer=tf.optimizers.Adam(learning_rate=0.01), loss=negloglik)
3   # Fit
4   model.fit(carat, price, epochs=100, verbose=0)
```

We can then obtain the model's predictions on the hold-out test data and visualize everything using our `plot_scatter()` function:

```
1   # Define range for model input
2   carat_hat = tf.linspace(carat_test.min(), carat_test.max(), 100)
3   # Obtain model's price predictions on test data
4   price_hat = model(carat_hat)
```

```
5  # Plot test data and model predictions
6  plot_scatter(carat_test, price_test, carat_hat, price_hat)
```

This produces the following chart:

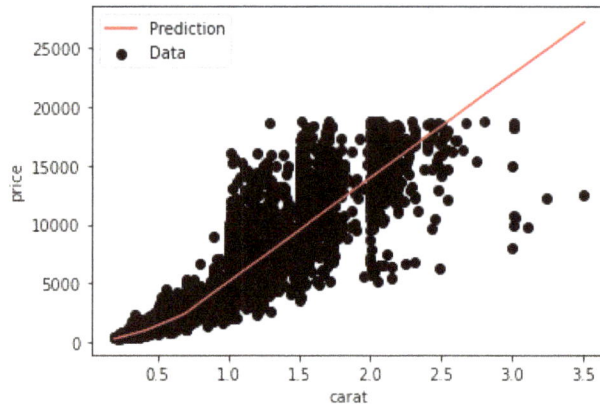

Figure 7.10: Predictions without uncertainty on diamond test data

We can see in *Figure 7.10* that the model captures the non-linear trend of the data. As a diamond's weight increases, the model predicts prices to increase more rapidly as we add more and more weight.

However, there is another obvious trend in the data that the model does not capture. We can observe that as weight increases, there is more and more variability in the price. At low weight, we only observe a little scatter around the fitted line, but the scatter increases at higher weight. We can consider this variability as inherent to the problem. That is, even if we had a lot more training data, we would still not be able to predict price, especially at high weight, perfectly. This sort of variability is aleatoric uncertainty, which we first encountered in *Chapter 4*, and will have a closer look at in the next subsection.

Step 3: Fitting a model with aleatoric uncertainty

We can account for aleatoric uncertainty in our model by predicting the standard deviation of the normal distribution in addition to predicting its mean. As before, we build a model with a normalizer layer and two dense layers. However, this time the second dense layer will output two values instead of one. The first output value will again be used to parameterize the mean of a normal distribution. But the second output value will parameterize the variance of the normal distribution, which allows us to quantify the aleatoric uncertainty in the model's predictions:

```
1   model_aleatoric = tf.keras.Sequential(
2       [
3           normalizer,
4           tf.keras.layers.Dense(32, activation="relu"),
5           tf.keras.layers.Dense(2),
6           tfp.layers.DistributionLambda(
7               lambda t: tfp.distributions.Normal(
8                   loc=t[..., :1], scale=1e-3 + tf.math.softplus(0.05 * t[..., 1:])
9               )
10          ),
11      ]
12  )
```

We again compile and fit the model on the weight and price data:

```
1   # Compile
2   model_aleatoric.compile(
3       optimizer=tf.optimizers.Adam(learning_rate=0.05), loss=negloglik
4   )
5   # Fit
6   model_aleatoric.fit(carat, price, epochs=100, verbose=0)
```

Now, we can obtain and visualize predictions on the test data. Note that this time, we pass `plot_std=True` in order to also plot the standard deviation of the predicted output distribution:

```
1  carat_hat = tf.linspace(carat_test.min(), carat_test.max(), 100)
2  price_hat = model_aleatoric(carat_hat)
3  plot_scatter(
4      carat_test, price_test, carat_hat, price_hat, plot_std=True,
5  )
```

We have now trained a model that represents the variation inherent to the data. The dashed error bars in *Figure 7.11* show the predicted variability of price as a function of weight. We can observe that the model is indeed less certain about what price to predict at weight above 1 carat, reflecting the larger scatter in the data that we observe at the higher weight range.

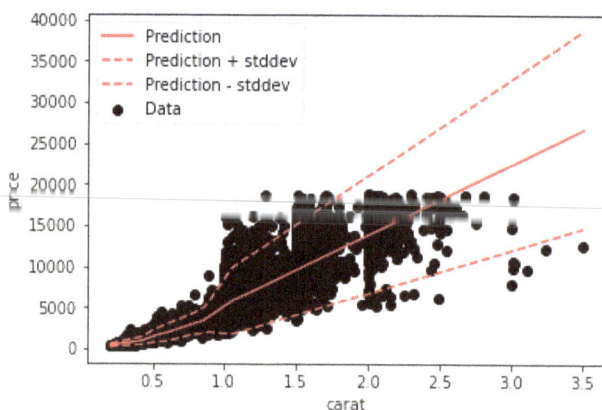

Figure 7.11: Predictions with aleatoric uncertainty on diamond test data

Step 4: Fitting a model with epistemic uncertainty

In addition to aleatoric uncertainty, we also deal with epistemic uncertainty – the uncertainty that stems not from the data, but from our model. Looking back at *Figure*

7.11, for example, the solid line, which represents the mean of our model prediction, appears to capture the trend of the data reasonably well. But given that training data is limited, we cannot be 100% certain that we found the true mean of the underlying data distribution. Maybe the true mean is actually a little bit greater or a little less than what we estimated it to be. In this section, we look at how we can model such uncertainty, and we will also see that epistemic uncertainty can be reduced by observing more data.

The trick to modeling epistemic uncertainty is, once again, to represent the weights in our neural network by a distribution rather than a point estimate. We can achieve this by replacing the dense layers that we used previously with DenseVariational layers from `tensorflow_probability`. Under the hood, this will implement BBB, which we first learned about in *Chapter 5*. In brief, when using BBB, we learn the posterior distribution over the weights of our network using the principle of variational learning. In order to do so, we need to define both prior and posterior distribution functions.

Note that the code example for BBB presented in *Chapter 5* made use of predefined `tensorflow_probability` modules for 2D convolution and dense layers with the reparameterization trick, which implicitly defined prior and posterior functions for us. In this example, we will define the prior and posterior functions for the dense layer ourselves.

We start by defining the prior over the dense layer's weights (both the kernel and the bias terms). The prior distribution models the uncertainty in the weights before we observe any data. It can be defined using a multivariate normal distribution that has a trainable mean and a variance that is fixed at 1:

```
1  def prior(kernel_size, bias_size=0, dtype=None):
2      n = kernel_size + bias_size
3      return tf.keras.Sequential(
4          [
5              tfp.layers.VariableLayer(n, dtype=dtype),
6              tfp.layers.DistributionLambda(
```

```
7                           lambda t: tfp.distributions.Independent(
8                                   tfp.distributions.Normal(loc=t, scale=1),
9                                   reinterpreted_batch_ndims=1,
10                              )
11                      ),
12              ]
13          )
```

We also define the variational posterior. The variational posterior is an approximation to the distribution of the dense layer's weights after we have observed the training data. We again use a multivariate normal distribution:

```
1   def posterior(kernel_size, bias_size=0, dtype=None):
2       n = kernel_size + bias_size
3       c = np.log(np.expm1(1.0))
4       return tf.keras.Sequential(
5           [
6                   tfp.layers.VariableLayer(2 * n, dtype=dtype),
7                   tfp.layers.DistributionLambda(
8                       lambda t: tfp.distributions.Independent(
9                           tfp.distributions.Normal(
10                              loc=t[..., :n],
11                              scale=1e-5 + tf.nn.softplus(c + t[..., n:]),
12                          ),
13                          reinterpreted_batch_ndims=1,
14                      )
15                  ),
16          ]
17      )
```

Equipped with these prior and posterior functions, we can define our model. As before, we use the normalizer layer to normalize our inputs and then stack two dense layers on top of each other. But this time, the dense layers will represent their parameters as distributions rather than point estimates. We achieve this by using the DenseVariational layers from `tensorflow_probability` together with our prior and posterior functions. The final output layer is a normal distribution with its variance set to 1 and with its mean parameterized by the output of the preceding DenseVariational layer:

```python
def build_epistemic_model():
    model = tf.keras.Sequential(
        [
            normalizer,
            tfp.layers.DenseVariational(
                32,
                make_prior_fn=prior,
                make_posterior_fn=posterior,
                kl_weight=1 / NUM_TRAIN_SAMPLES,
                activation="relu",
            ),
            tfp.layers.DenseVariational(
                1,
                make_prior_fn=prior,
                make_posterior_fn=posterior,
                kl_weight=1 / NUM_TRAIN_SAMPLES,
            ),
            tfp.layers.DistributionLambda(
                lambda t: tfp.distributions.Normal(loc=t, scale=1)
            ),
        ]
    )
    return model
```

To observe the effect of the amount of available training data on epistemic uncertainty estimates, we first fit our model on a small subset of data before fitting it on all available training data. We take the first 500 samples from the training dataset:

```
1  carat_subset = carat[:500]
2  price_subset = price[:500]
```

We build, compile, and fit the model as before:

```
1  # Build
2  model_epistemic = build_epistemic_model()
3  # Compile
4  model_epistemic.compile(
5      optimizer=tf.optimizers.Adam(learning_rate=0.01), loss=negloglik
6  )
7  # Fit
8  model_epistemic.fit(carat_subset, price_subset, epochs=100, verbose=0)
```

We then obtain and plot predictions on the test data. Note that here, we sample from the posterior distribution 10 times, which allows us to observe how much the predicted mean varies with every sample iteration. If the predicted mean varies a lot, this means that epistemic uncertainty is estimated to be large, while if the mean varies only very little, there is only little estimated epistemic uncertainty:

```
1  carat_hat = tf.linspace(carat_test.min(), carat_test.max(), 100)
2  price_hats = [model_epistemic(carat_hat) for _ in range(10)]
3  plot_scatter(
4      carat_test, price_test, carat_hat, price_hats,
5  )
```

In *Figure 7.12*, we can observe that the predicted mean varies over the 10 different samples. Interestingly, variation (and thus epistemic uncertainty) seems to be lower at lower weights and increases as weight increases.

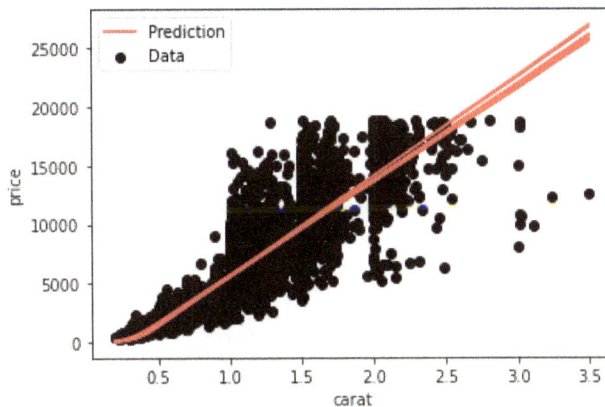

Figure 7.12: Predictions with high epistemic uncertainty on diamond test data

In order to verify that epistemic uncertainty can be reduced by training on more data, we train our model on the full training dataset:

```
1  # Build
2  model_epistemic_full = build_epistemic_model()
3  # Compile
4  model_epistemic_full.compile(
5      optimizer=tf.optimizers.Adam(learning_rate=0.01), loss=negloglik
6  )
7  # Fit
8  model_epistemic_full.fit(carat, price, epochs=100, verbose=0)
```

And then plot the predictions for the full data model:

```
1  carat_hat = tf.linspace(carat_test.min(), carat_test.max(), 100)
2  price_hats = [model_epistemic_full(carat_hat) for _ in range(10)]
3  plot_scatter(
4      carat_test, price_test, carat_hat, price_hats,
5  )
```

As expected, we see in *Figure 7.13* that epistemic uncertainty is much lower now and the predicted mean varies very little over the 10 samples (to the point where it is hard to see any difference between the 10 red curves):

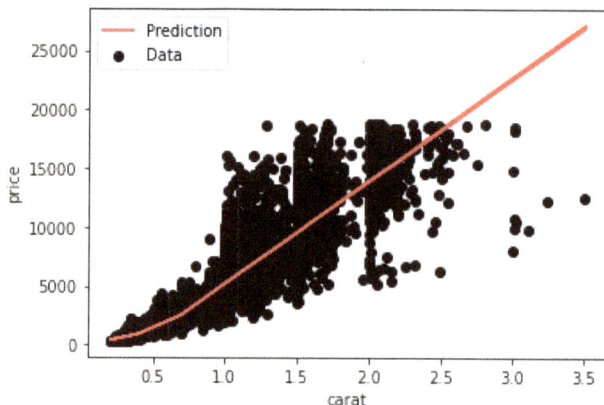

Figure 7.13: Predictions with low epistemic uncertainty on diamond test data

Step 5: Fitting a model with aleatoric and epistemic uncertainty

As a final exercise, we can put all the building blocks together and build a neural network that models both aleatoric and epistemic uncertainty. We can achieve this by using two DenseVariational layers (which will allow us to model epistemic uncertainty) and then stacking a normal distribution layer on top whose mean and variance are parameterized by the outputs of the second DenseVariational layer (which will allow us to model aleatoric uncertainty):

```
1   # Build model.
2   model_epistemic_aleatoric = tf.keras.Sequential(
3       [
4           normalizer,
5           tfp.layers.DenseVariational(
6               32,
7               make_prior_fn=prior,
```

```
 8              make_posterior_fn=posterior,
 9              kl_weight=1 / NUM_TRAIN_SAMPLES,
10              activation="relu",
11          ),
12          tfp.layers.DenseVariational(
13              1 + 1,
14              make_prior_fn=prior,
15              make_posterior_fn=posterior,
16              kl_weight=1 / NUM_TRAIN_SAMPLES,
17          ),
18          tfp.layers.DistributionLambda(
19              lambda t: tfp.distributions.Normal(
20                  loc=t[..., :1], scale=1e-3 + tf.math.softplus(0.05 * t[..., 1:])
21              )
22          ),
23      ]
24  )
```

We can build and train this model following the same procedure as before. We can then again perform inference on the test data for 10 times, which yields the predictions shown in *Figure 7.14*. Every of the 10 inferences now yields a predicted mean and standard deviation. The standard deviation represents the estimated aleatoric uncertainty for every inference, and the variation observed across the different inferences represents the epistemic uncertainty.

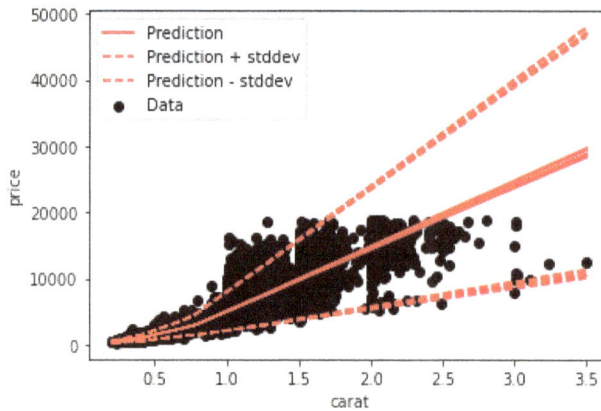

Figure 7.14: Predictions with both epistemic and aleatoric uncertainty on diamond test data

7.3.1 Sources of uncertainty: Image classification case study

In the previous case study, we saw how we can model aleatoric and epistemic uncertainty in a regression problem. In this section, we'll look at the MNIST digits dataset one more time to model aleatoric and epistemic uncertainty. We will also explore how aleatoric uncertainty can be difficult to reduce, whereas epistemic uncertainty can be reduced with more data.

Let's start with our data. To make our example more insightful, we will not just use the standard MNIST dataset but also use a variant of MNIST named AmbiguousMNIST. This dataset contains generated images that are, unsurprisingly, inherently ambiguous. Let's first load the data and then explore the AmbiguousMNIST dataset. We'll start with the necessary imports:

```
1  import tensorflow as tf
2  import tensorflow_probability as tfp
3  import matplotlib.pyplot as plt
4  import numpy as np
5  from sklearn.utils import shuffle
```

```
6   from sklearn.metrics import roc_auc_score
7   import ddu_dirty_mnist
8   from scipy.stats import entropy
9   tfd = tfp.distributions
```

We can download the AmbiguousMNIST dataset with the ddu_dirty_mnist library:

```
1   dirty_mnist_train = ddu_dirty_mnist.DirtyMNIST(
2       ".",
3       train=True,
4       download=True,
5       normalize=False,
6       noise_stddev=0
7   )
8
9   # regular MNIST
10  train_imgs = dirty_mnist_train.datasets[0].data.numpy()
11  train_labels = dirty_mnist_train.datasets[0].targets.numpy()
12  # AmbiguousMNIST
13  train_imgs_amb = dirty_mnist_train.datasets[1].data.numpy()
14  train_labels_amb = dirty_mnist_train.datasets[1].targets.numpy()
```

We then concatenate and shuffle the images and labels so that we have a good mix of both datasets during training. We also fix the shape of the dataset so that it fits the setup of our model:

```
1   train_imgs, train_labels = shuffle(
2       np.concatenate([train_imgs, train_imgs_amb]),
3       np.concatenate([train_labels, train_labels_amb])
4   )
5   train_imgs = np.expand_dims(train_imgs[:, 0, :, :], -1)
6   train_labels = tf.one_hot(train_labels, 10)
```

Figure 7.15 gives an example of the AmbiguousMNIST images. We can see that the images are in between classes: a 4 can also be interpreted as an 9, a 0 can be interpreted as a 6, and vice versa. This means that our model will most likely struggle to classify at least a portion of these images correctly as they are inherently noisy.

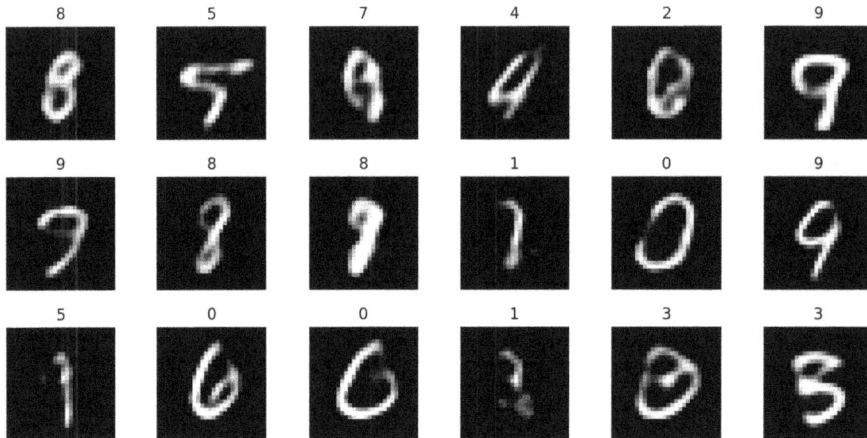

Figure 7.15: Examples of images from the AmbiguousMNIST dataset

Now that we have our train dataset, let's load our test dataset as well. We will just use the standard MNIST test dataset:

```
1  (test_imgs, test_labels) = tf.keras.datasets.mnist.load_data()[1]
2  test_imgs = test_imgs / 255.
3  test_imgs = np.expand_dims(test_imgs, -1)
4  test_labels = tf.one_hot(test_labels, 10)
```

We can now start to define our model. In this example, we use a small Bayesian neural net with **Flipout** layers. These layers sample from the kernel and bias posteriors during the forward pass and thus add stochasticity to our model. We can use this later on when we want to compute uncertainty values:

```
1  kl_divergence_function = lambda q, p, _: tfd.kl_divergence(q, p) / tf.cast(
2      60000, dtype=tf.float32
3  )
4
5  model = tf.keras.models.Sequential(
6      [
7          *block(5),
8          *block(16),
9          *block(120, max_pool=False),
10         tf.keras.layers.Flatten(),
11         tfp.layers.DenseFlipout(
12             84,
13             kernel_divergence_fn=kl_divergence_function,
14             activation=tf.nn.relu,
15         ),
16         tfp.layers.DenseFlipout(
17             10,
18             kernel_divergence_fn=kl_divergence_function,
19             activation=tf.nn.softmax,
20         ),
21     ]
22 )
```

We define a block as follows:

```
1  def block(filters: int, max_pool: bool = True):
2      conv_layer =  tfp.layers.Convolution2DFlipout(
3              filters,
4              kernel_size=5,
5              padding="same",
6              kernel_divergence_fn=kl_divergence_function,
7              activation=tf.nn.relu)
```

```
 8          if not max_pool:
 9              return (conv_layer,)
10          max_pool = tf.keras.layers.MaxPooling2D(
11              pool_size=[2, 2], strides=[2, 2], padding="same"
12          )
13          return conv_layer, max_pool
```

We compile our model and can start training:

```
 1  model.compile(
 2          tf.keras.optimizers.Adam(),
 3          loss="categorical_crossentropy",
 4          metrics=["accuracy"],
 5          experimental_run_tf_function=False,
 6  )
 7  model.fit(
 8          x=train_imgs,
 9          y=train_labels,
10          validation_data=(test_imgs, test_labels),
11          epochs=50
12  )
```

We are now interested in separating images via epistemic uncertainty and aleatoric uncertainty. Epistemic uncertainty should separate our in-distribution images from out-of-distribution images, as these images can be seen as unknown unknowns: our model has never seen these images before, and should therefore assign high epistemic uncertainty (or *knowledge uncertainty*) to them. Although our model was trained on the AmbiguousMNIST dataset, the model should still have high aleatoric uncertainty when it would see images from this dataset at test time: training with these images does not reduce aleatoric uncertainty (or *data uncertainty*) as the images are inherently ambiguous.

We use the FashionMNIST dataset as the out-of-distribution dataset. We use the
AmbiguousMNIST test set as our ambiguous dataset for testing:

```
1  (_, _), (ood_imgs, _) = tf.keras.datasets.fashion_mnist.load_data()
2  ood_imgs = np.expand_dims(ood_imgs / 255., -1)
3
4  ambiguous_mnist_test = ddu_dirty_mnist.AmbiguousMNIST(
5      ".",
6      train=False,
7      download=True,
8      normalize=False,
9      noise_stddev=0
10 )
11 amb_imgs = ambiguous_mnist_test.data.numpy().reshape(60000, 28, 28, 1)[:10000]
12 amb_labels = tf.one_hot(ambiguous_mnist_test.targets.numpy(), 10).numpy()
```

Let's use the stochasticity of our model to create a variety of model predictions. We
iterate over our test images fifty times:

```
1  preds_id = []
2  preds_ood = []
3  preds_amb = []
4  for _ in range(50):
5      preds_id.append(model(test_imgs))
6      preds_ood.append(model(ood_imgs))
7      preds_amb.append(model(amb_imgs))
8  # format data such that we have it in shape n_images, n_predictions, n_classes
9  preds_id = np.moveaxis(np.stack(preds_id), 0, 1)
10 preds_ood = np.moveaxis(np.stack(preds_ood), 0, 1)
11 preds_amb = np.moveaxis(np.stack(preds_amb), 0, 1)
```

We can then define some functions to compute the different kinds of uncertainty:

```python
def total_uncertainty(preds: np.ndarray) -> np.ndarray:
    return entropy(np.mean(preds, axis=1), axis=-1)

def data_uncertainty(preds: np.ndarray) -> np.ndarray:
    return np.mean(entropy(preds, axis=2), axis=-1)

def knowledge_uncertainty(preds: np.ndarray) -> np.ndarray:
    return total_uncertainty(preds) - data_uncertainty(preds)
```

Finally, we can see how well our model can distinguish between in-distribution, ambiguous, and out-of-distribution images. Let's plot a histogram of the different distributions according to the different uncertainty methods:

```python
labels = ["In-distribution", "Out-of-distribution", "Ambiguous"]
uncertainty_functions = [total_uncertainty, data_uncertainty, knowledge_uncertainty]
fig, axes = plt.subplots(1, 3, figsize=(20,5))
for ax, uncertainty in zip(axes, uncertainty_functions):
    for scores, label in zip([preds_id, preds_ood, preds_amb], labels):
        ax.hist(uncertainty(scores), bins=20, label=label, alpha=.8)
    ax.title.set_text(uncertainty.__name__.replace("_", " ").capitalize())
    ax.legend(loc="upper right")
plt.legend()
plt.savefig("uncertainty_types.png", dpi=300)
plt.show()
```

This produces the following output:

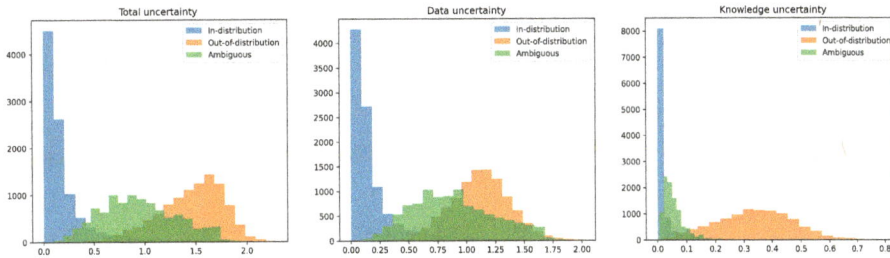

Figure 7.16: The different types of uncertainty on MNIST

What can we observe?

- Total and data uncertainty are relatively good at distinguishing in-distribution data from out-of-distribution and ambiguous data.

- However, data and total uncertainty are not able to separate ambiguous data from out-of-distribution data. To do that, we need knowledge uncertainty. We can see that knowledge uncertainty clearly separates ambiguous data from out-of-distribution data.

- We trained on ambiguous samples as well, but that doesn't reduce the uncertainty of the ambiguous test samples to uncertainty levels similar to the original in-distribution data. This shows that data uncertainty cannot easily be reduced. The data is inherently ambiguous, no matter how much ambiguous data the model sees.

We can confirm these observations by looking at the AUROC for the different combinations of distributions.

We can first compute the AUROC score to compute the ability of our model to separate in-distribution and ambiguous images from out-of-distribution images:

```
1  def auc_id_and_amb_vs_ood(uncertainty):
2      scores_id = uncertainty(preds_id)
```

```
3        scores_ood = uncertainty(preds_ood)
4        scores_amb = uncertainty(preds_amb)
5        scores_id = np.concatenate([scores_id, scores_amb])
6        labels = np.concatenate([np.zeros_like(scores_id), np.ones_like(scores_ood)])
7        return roc_auc_score(labels, np.concatenate([scores_id, scores_ood]))
8
9
10  print(f"{auc_id_and_amb_vs_ood(total_uncertainty)=:.2%}")
11  print(f"{auc_id_and_amb_vs_ood(knowledge_uncertainty)=:.2%}")
12  print(f"{auc_id_and_amb_vs_ood(data_uncertainty)=:.2%}")
13  # output:
14  # auc_id_and_amb_vs_ood(total_uncertainty)=91.81%
15  # auc_id_and_amb_vs_ood(knowledge_uncertainty)=98.87%
16  # auc_id_and_amb_vs_ood(data_uncertainty)=84.29%
```

We see a confirmation of what we saw in our histograms: knowledge uncertainty is far better than the other two types of uncertainty at separating in-distribution and ambiguous data from out-of-distribution data.

```
1  def auc_id_vs_amb(uncertainty):
2      scores_id, scores_amb = uncertainty(preds_id), uncertainty(preds_amb)
3      labels = np.concatenate([np.zeros_like(scores_id), np.ones_like(scores_amb)])
4      return roc_auc_score(labels, np.concatenate([scores_id, scores_amb]))
5
6
7  print(f"{auc_id_vs_amb(total_uncertainty)=:.2%}")
8  print(f"{auc_id_vs_amb(knowledge_uncertainty)=:.2%}")
9  print(f"{auc_id_vs_amb(data_uncertainty)=:.2%}")
10  # output:
11  # auc_id_vs_amb(total_uncertainty)=94.71%
12  # auc_id_vs_amb(knowledge_uncertainty)=87.06%
13  # auc_id_vs_amb(data_uncertainty)=95.21%
```

We can see that both total uncertainty and data uncertainty are able to separate in-distribution from ambiguous data pretty well. Using data uncertainty gives us a small improvement over using total uncertainty. Knowledge uncertainty, however, is not able to distinguish between in-distribution data and ambiguous data.

7.4 Summary

In this chapter, we've taken a look at a number of practical considerations of using Bayesian deep learning: exploring trade-offs in model performance and learning how we can use Bayesian neural network methods to better understand the effects of different uncertainty sources on our data.

In the next chapter, we'll dig further into applying BDL through a variety of case studies, demonstrating the benefits of these methods in a range of practical settings.

7.5 Further reading

- *Practical Considerations for Probabilistic Backpropagation*, Matt Benatan *et al.*: In this paper, the authors explore methods to get the most out of PBP, demonstrating how different early stopping approaches can be used to improve training, exploring the tradeoffs associated with mini-batching, and more

- *Modeling aleatoric and epistemic uncertainty using TensorFlow and TensorFlow Probability*, Alexander Molak: In this Jupyter notebook, the author shows how to model aleatoric and epistemic uncertainty on regression toy data

- *Weight Uncertainty in Neural Networks*, Charles Blundell *et al.*: In this paper, the authors introduce BBB, which we use in the regression case study and is one of the key pieces of BDL literature

- *Deep Deterministic Uncertainty: A Simple Baseline*, Jishnu Mukhoti *et al.*: In this work, the authors describe several experiments related to the different types of uncertainty and introduce the *AmbiguousMNIST* dataset that we used in the last case study

- *Uncertainty Estimation in Deep Learning with application to Spoken Language Assessment*, Andrey Malinin: This thesis highlights the different sources of uncertainty with intuitive examples

8

Applying Bayesian Deep Learning

This chapter will guide you through a variety of applications of Bayesian deep learning (BDL). These will include the use of BDL in standard classification tasks, as well as demonstrating how it can be used in more sophisticated ways for out-of-distribution detection, data selection, and reinforcement learning.

We will cover these topics in the following sections:

- Detecting out-of-distribution data

- Being robust against dataset drift

- Using data selection via uncertainty to keep models fresh

- Using uncertainty estimates for smarter reinforcement learning

- Susceptibility to adversarial input

8.1   Technical requirements

All of the code for this book can be found on the GitHub repository for the book: `https://github.com/PacktPublishing/Enhancing-Deep-Learning-with-Bayesian-Inference`.

8.2   Detecting out-of-distribution data

Typical neural networks do not handle out-of-distribution data well. We saw in *Chapter 3* that a cat-dog classifier classified an image of a parachute as a dog with more than 99% confidence. In this section, we will look into what we can do about this vulnerability of neural networks. We will do the following:

- Explore the problem visually by perturbing a digit of the `MNIST` dataset

- Explain the typical way out-of-distribution detection performance is reported in the literature

- Review the out-of-distribution detection performance of some of the standard practical BDL methods we look at in this chapter

- Explore even more practical methods that are specifically tailored to detect out-of-distribution detection

8.2.1   Exploring the problem of out-of-distribution detection

To give you a better understanding of what out-of-distribution performance is like, we will start with a visual example. Here is what we will do:

- We will train a standard network on the `MNIST` digit dataset

- We will then perturb a digit and gradually make it more out-of-distribution

- We will report the confidence score of a standard model and MC dropout

With this visual example, we can see how simple Bayesian methods can improve the out-of-distribution detection performance over a standard deep learning model. We start by training a simple model on the `MNIST` dataset.

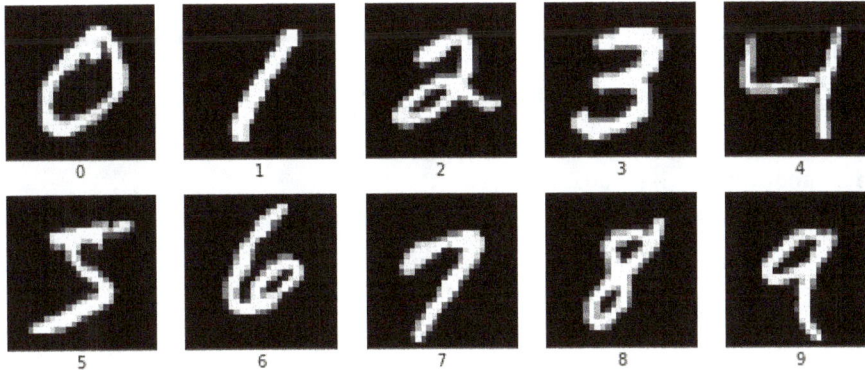

Figure 8.1: The classes of the MNIST dataset: 28x28 pixel images of the digits zero to nine

We use `TensorFlow` to train our model, `numpy` to make our images more out-of-distribution, and `Matplotlib` to visualize our data.

```
1  import tensorflow as tf
2  from tensorflow.keras import datasets, layers, models
3  import numpy as np
4  import matplotlib.pyplot as plt
```

The `MNIST` dataset is available in TensorFlow, so we can just load it:

```
1  (train_images, train_labels), (
2      test_images,
3      test_labels,
4  ) = datasets.mnist.load_data()
5  train_images, test_images = train_images / 255.0, test_images / 255.0
```

MNIST is a simple dataset, so a simple model allows us to achieve a test accuracy of more than 99%. We use a standard CNN with three convolutional layers:

```python
def get_model():
    model = models.Sequential()
    model.add(
        layers.Conv2D(32, (3, 3), activation="relu", input_shape=(28, 28, 1))
    )
    model.add(layers.MaxPooling2D((2, 2)))
    model.add(layers.Conv2D(64, (3, 3), activation="relu"))
    model.add(layers.MaxPooling2D((2, 2)))
    model.add(layers.Conv2D(64, (3, 3), activation="relu"))
    model.add(layers.Flatten())
    model.add(layers.Dense(64, activation="relu"))
    model.add(layers.Dense(10))
    return model


model = get_model()
```

We can then compile and train our model. We obtain a validation accuracy of over 99% after just 5 epochs

```python
def fit_model(model):
    model.compile(
        optimizer="adam",
        loss=tf.keras.losses.SparseCategoricalCrossentropy(from_logits=True),
        metrics=["accuracy"],
    )

    model.fit(
        train_images,
```

```
10              train_labels,
11              epochs=5,
12              validation_data=(test_images, test_labels),
13          )
14      return model
15
16
17  model = fit_model(model)
```

Now, let's see how this model handles out-of-distribution data. Imagine we deploy this model to recognize digits, but users sometimes fail to write down the entire digit. What happens when users do not write down the entire digit? We can get an answer to this question by gradually removing more and more information from a digit, and seeing how our model handles the perturbed inputs. We can define our function to remove signal as follows:

```
1  def remove_signal(img: np.ndarray, num_lines: int) -> np.ndarray:
2      img = img.copy()
3      img[:num_lines] = 0
4      return img
```

And then we perturb our images:

```
1  imgs = []
2  for i in range(28):
3      img_perturbed = remove_signal(img, i)
4      if np.array_equal(img, img_perturbed):
5          continue
6      imgs.append(img_perturbed)
7      if img_perturbed.sum() == 0:
8          break
```

We only add perturbed images to our list of images if setting a row to 0 actually changes the original image (`if` `np.array_equal(img, img_perturbed)`) and stop once the image is completely black, meaning it just contains pixels with a value of 0. We run inference on these images:

```
1  softmax_predictions = tf.nn.softmax(model(np.expand_dims(imgs, -1)), axis=1)
```

We can then plot all images with their predicted labels and confidence scores:

```
1  plt.figure(figsize=(10, 10))
2  bbox_dict = dict(
3      fill=True, facecolor="white", alpha=0.5, edgecolor="white", linewidth=0
4  )
5  for i in range(len(imgs)):
6      plt.subplot(5, 5, i + 1)
7      plt.xticks([])
8      plt.yticks([])
9      plt.grid(False)
10     plt.imshow(imgs[i], cmap="gray")
11     prediction = softmax_predictions[i].numpy().max()
12     label = np.argmax(softmax_predictions[i])
13     plt.xlabel(f"{label} - {prediction:.2%}")
14     plt.text(0, 3, f" {i+1}", bbox=bbox_dict)
15 plt.show()
```

This produces the following figure:

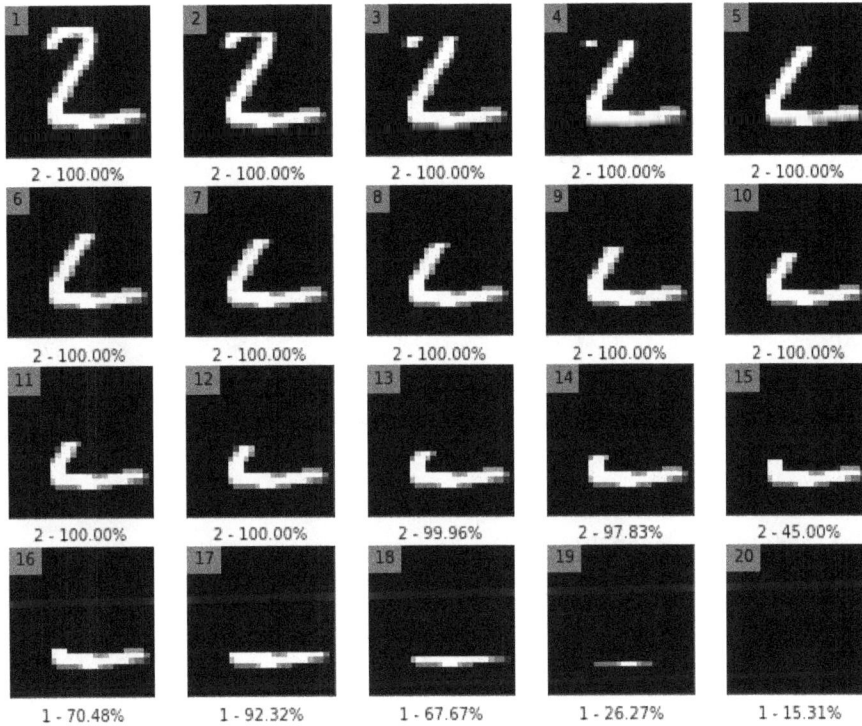

Figure 8.2: Predicted label and corresponding softmax score of a standard neural network for an image that is more and more out-of-distribution

We can see in *Figure 8.2* that, initially, our model confidently classifies the image as a **2**. Remarkably, this confidence persists even when it seems unreasonable to do so. For example, the model still classifies image 14 as a **2** with 97.83% confidence. Moreover, the model predicts with 92.32% confidence that a completely horizontal line is a **1**, as we can see in image 17. It looks like our model is overconfident in its predictions.

Let's see what a slightly different model would predict on these images. We'll now use MC dropout as our model. By sampling, we should be able to increase the models' uncertainty compared to a standard NN. Let's first define our model:

```python
def get_dropout_model():
    model = models.Sequential()
    model.add(
        layers.Conv2D(32, (3, 3), activation="relu", input_shape=(28, 28, 1))
    )
    model.add(layers.Dropout(0.2))
    model.add(layers.MaxPooling2D((2, 2)))
    model.add(layers.Conv2D(64, (3, 3), activation="relu"))
    model.add(layers.MaxPooling2D((2, 2)))
    model.add(layers.Dropout(0.5))
    model.add(layers.Conv2D(64, (3, 3), activation="relu"))
    model.add(layers.Dropout(0.5))
    model.add(layers.Flatten())
    model.add(layers.Dense(64, activation="relu"))
    model.add(layers.Dropout(0.5))
    model.add(layers.Dense(10))
    return model
```

Then let's instantiate it:

```python
dropout_model = get_dropout_model()
dropout_model = fit_model(dropout_model)
```

Our model with dropout will achieve a similar accuracy as our vanilla model. Let's now run inference with dropout and plot the mean confidence score of MC dropout:

```
1  predictions = np.array(
2      [
3          tf.nn.softmax(dropout_model(imgs_np, training=True), axis=1)
4          for _ in range(100)
5      ]
6  )
7  predictions_mean = np.mean(predictions, axis=0)
8  plot_predictions(predictions_mean)
```

This again produces a figure showing the predicted labels and their associated confidence scores:

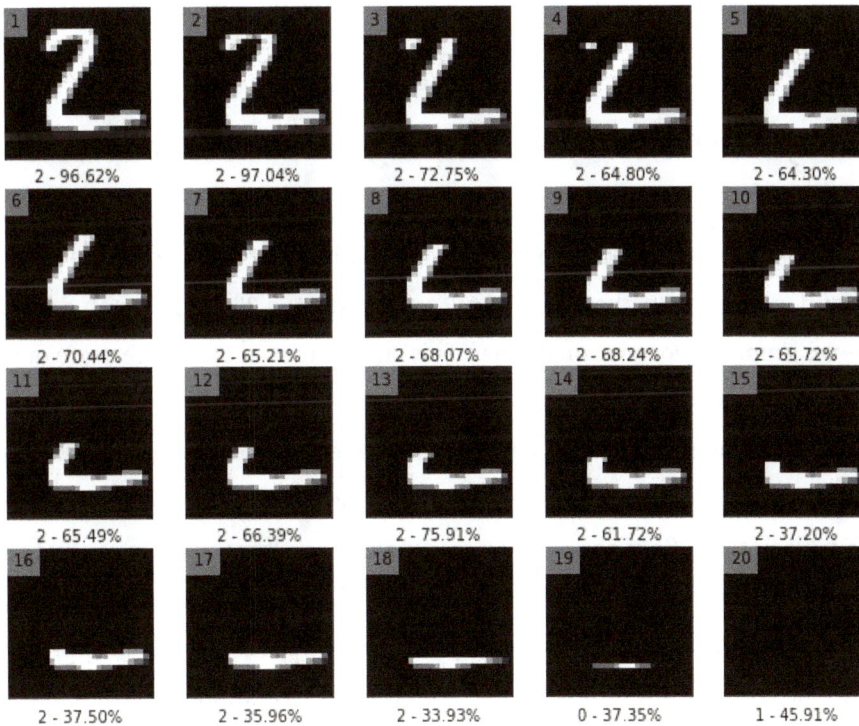

Figure 8.3: Predicted label and corresponding softmax score of an MC dropout network for an image that is more and more out-of-distribution

We can see in *Figure 8.3* that the model is less certain on average. The model's confidence decreases a lot when we remove rows from our image. That is desired behaviour: our model does not know the input, so it should be uncertain. However, we can also see that the model is not perfect:

- It maintains a pretty high confidence for images that do not really look like a **2**.

- The model's confidence can change a lot when we delete one more row from our images. For example, the model's confidence jumps from 61.72% to 37.20% between image 14 and 15.

- The model seems to be more confident that image 20, without any white pixels, is a **1**.

MC dropout is, in this case, a step in the right direction, but is not handling the out-of-distribution data perfectly.

8.2.2 Systematically evaluating OOD detection performance

The preceding example suggests that MC dropout gives out-of-distribution images a lower confidence score on average. But we only evaluated 20 images with a limited variety – we simply removed a single row. This change moved the image more out-of-distribution, but all images shown in the previous section are relatively similar to the training distribution of MNIST if you compare it to, let's say, natural images of objects. Images of airplanes, cars, or birds will definitely be much more out-of-distribution than an image of MNIST with a few black rows. So it seems to be reasonable that, if we want to evaluate the OOD detection performance of our model, we should also test it on images that are even more OOD, that is, from a completely different dataset. This is the approach that is typically taken in the literature to evaluate out-of-distribution detection performance. The procedure is as follows:

1. We train a model on in-distribution (ID) images.

2. We take one or more completely different OOD datasets and feed these to our model.

3. We now treat the predictions of the model on the ID and OOD test datasets as a binary problem and compute a single score for every image.

   - In the case of evaluation of the softmax score, this means that we take the model's maximum softmax score for every ID and OOD image.

4. With these scores, we can compute binary metrics, such as the area under the receiver operating characteristic (AUROC).

The better the model performs on these binary metrics, the better the model's OOD detection performance.

8.2.3   Simple out-of-distribution detection without retraining

Although MC dropout can be an effective method to detect out-of-distribution data, it comes with a major disadvantage at inference time: we need to run inference five, or maybe even a hundred, times instead of just once. Something similar can be said for certain other Bayesian deep learning methods: although they are principled, they are not always the most practical way to obtain a good OOD detection performance. The main downside is that they often require retraining of your network, which can be expensive to do if you have a lot of data. This is why there is an entire field of OOD detection methods that are not explicitly grounded on Bayesian theory, but can provide a good, simple, or even excellent baseline. These methods often do not require any retraining and can be applied out of the box on a standard neural network. Two methods that are often used in the OOD detection literature are worth mentioning:

- **ODIN**: OOD detection with preprocessing and scaling

- **Mahalanobis**: OOD detection with intermediate features

ODIN: OOD detection with preprocessing and scaling

Out-of-DIstribution detector for Neural networks (ODIN) is one of the standard methods in practical out-of-distribution detection because of its simplicity and effectiveness. Although the method was introduced in 2017, it is still frequently used as a comparison method in papers that propose out-of-distribution detection methods.

ODIN consists of two key ideas:

- **Temperature scaling** of the logit scores before applying the softmax operation to improve the ability of the softmax score to distinguish between in- and out-of-distribution images

- **Input preprocessing** to make in-distribution images more in-distribution

Let's look at both ideas in a bit more detail.

Temperature scaling

ODIN works for classification models. Given our softmax score computed as

$$p_i(\boldsymbol{x}) = \frac{\exp\left(f_i(\boldsymbol{x})\right)}{\sum_{j=1}^{N} \exp\left(f_j(\boldsymbol{x})\right)} \tag{8.1}$$

Here, $f_i(x)$ is a single logit output and $f_j(x)$ are the logits for all classes for a single example, temperature scaling means that we divide these logit outputs by a constant T:

$$p_i(\boldsymbol{x}; T) = \frac{\exp\left(f_i(\boldsymbol{x})/T\right)}{\sum_{j=1}^{N} \exp\left(f_j(\boldsymbol{x})/T\right)} \tag{8.2}$$

For large values of T, temperature scaling causes the softmax scores to be closer to a uniform distribution, which helps to reduce overconfident predictions.

We can apply temperature scaling in Python, given a simple model that outputs the logits:

```
1  logits = model.predict(images)
2  logits_scaled = logits / temperature
3  softmax = tf.nn.softmax(logits, axis=1)
```

Input preprocessing

We saw in *Chapter 3* that the **Fast-Gradient Sign Method (FGSM)** allowed us to fool a neural network. By slightly changing an image of a cat, we could make the model predict "dog" with 99.41% confidence. The idea here was that we could take the sign of the gradient of the loss with respect to the input, multiply it by a small value and add that noise to our image – this moved our image away from our in-distribution class. By doing the opposite, that is, subtracting the noise from our image, we make the image more in-distribution. The authors of the ODIN paper show that this causes in-distribution images to have an even higher softmax score compared to out-of-distribution images. This means that we increase the difference between OOD and ID softmax scores, leading to a better OOD detection performance.

$$\tilde{x} = x - \varepsilon \operatorname{sign} \left(-\nabla_x \log S_{\hat{y}}(x; T) \right) \tag{8.3}$$

Where x is an input image of which we subtract the perturbation magnitude ϵ times the sign of the gradient of the cross-entropy loss with respect to the input. See *Chapter 3* for the TensorFlow implementation of this technique.

Although input preprocessing and temperature scaling are simple to implement, ODIN now requires two more hyperparameters to be tuned: the temperature for scaling the logits and ϵ of the inverse of the fast gradient sign method. ODIN uses a separate out-of-distribution dataset to tune these hyperparameters (the validation set of the iSUN dataset: 8925 images).

Mahalanobis: OOD Detection with intermediate features

In *A Simple Unified Framework for Detecting Out-of-Distribution Samples and Adversarial Attacks*, Kimin Lee et al. propose a different method to detect OOD input. The core of their method is the idea that each class of a classifier follows a multivariate Gaussian distribution in the feature space of a network. Given this idea, we can define C class-conditional Gaussian distributions with a tied covariance σ:

$$P(f(x) \mid y = c) = \mathcal{N}\left(f(x) \mid \mu_c, \sigma\right) \tag{8.4}$$

Where μ_c is the mean of the multivariate Gaussian distribution for each class c. This allows us to compute the empirical mean and covariance of each of these distributions for a given output of an intermediate layer, one for each class of our network. Based on the mean and covariance, we can compute the Mahalanobis distance of a single test image compared to our in-distribution data. We compute this for the class that is closest to the input image:

$$M(x) = \max_c - \left(f(x) - \widehat{\mu}_c\right)^\top \widehat{\sigma}^{-1} \left(f(x) - \widehat{\mu}_c\right) \tag{8.5}$$

This distance should be small for in-distribution images and large for out-of-distribution images.

numpy has convenient functions to compute the mean and covariance of an array:

```
1  mean = np.mean(features_of_class, axis=0)
2  covariance = np.cov(features_of_class.T)
```

Given these, we can compute the Mahalanobis distance as such:

```
1  covariance_inverse = np.linalg.pinv(covariance)
2  x_minus_mu = features_of_class - mean
```

```
3  mahalanobis = np.dot(x_minus_mu, covariance_inverse).dot(x_minus_mu.T)
4  mahalanobis = np.sqrt(mahalanobis).diagonal()
```

The Mahalanobis distance computation does not require any retraining and is a relatively cheap operation to perform once you have stored the mean and (inverse of the) covariance of the classes for the features of a layer of your network.

To improve the performance of the method, the authors show that we can also apply the input preprocessing as mentioned in the ODIN paper, or compute and then average the Mahalanobis distances extracted from multiple layers of the network.

8.3   Being robust against dataset shift

We already encountered dataset shift in *Chapter 3*. As a reminder, dataset shift is a common problem in machine learning that happens when the joint distribution $P(X, Y)$ of inputs X and outputs Y differs between the model training stage and model inference stage (for example, when testing the model or when running it in a production environment). Covariate shift is a specific case of dataset shift where only the distribution of the inputs changes but the conditional distribution $P(Y|X)$ stays constant.

Dataset shift is present in most production environments because of the difficulty of including all possible inference conditions during training and because most data is not static but changes over time. The input data can shift along many different dimensions in a production environment. Geographic and temporal dataset shift are two common forms of shift. Imagine, for example, you have trained your model on data taken from one geographical region (for example, Europe) and then apply the model in a different geographical region (for example, Latin America). Similarly, a model could be trained on data from the years between 2010 and 2020 and then applied on production data taken from today.

We will see that in such data shift scenarios, models often perform worse on the new shifted data than on their original training distribution. We will also see how vanilla neural networks usually do not indicate when the input data deviates from the training distribution. Finally, we will explore how various methods introduced in this book can be used to indicate dataset shift via uncertainty estimates and how these methods can make the models more robust. The following code example will be focused on an image classification problem. It should be noted, however, that the insights tend to generalize to other domains (such as natural language processing) and tasks (such as regression).

8.3.1   Measuring a model's response to dataset shift

Assuming that we have a training dataset and a separate test set, how can we measure a model's ability to signal to us when the data has shifted? In order to do so, it would be necessary to have an additional test set where the data has been shifted to check how the model reacts to the dataset shift. One commonly applied way to create such a data shift test set for images was originally suggested by Dan Hendrycks and Thomas Dietterich in 2019 and others. The idea is straightforward: take the images from your initial test set, then apply different image quality corruptions at different severity levels to them. Hendrycks and Dietterich proposed a set of 15 different types of image quality corruptions, ranging from image noise, blur, weather corruptions (such as fog and snow), and digital corruption. Each corruption type has five levels of severity, ranging from 1 (mild corruption) to 5 (severe corruption). *Figure 8.4* shows what the image of a kitty looks like initially (left) and after applying shot noise corruption to the image, either at severity level 1 (middle) or severity level 5 (right).

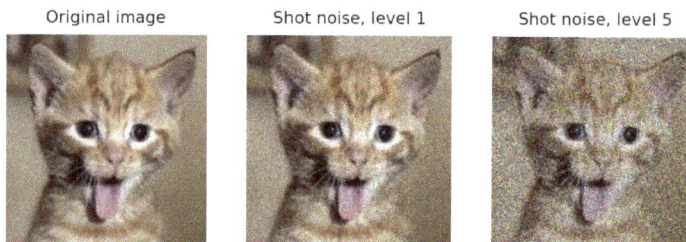

Figure 8.4: Generating artificial dataset shift by applying image quality corruptions at different levels of corruption severity

All these image quality corruptions can be generated conveniently using the `imgaug`
Python package. The following code assumes that we have an image called "kitty.png"
on disk. We load the image using the PIL package. We then specify the corruption
type (for example, `ShotNoise`) via the name of the corruption function, and then apply
the corruption function to the image, using either severity level 1 or 5 by passing the
corresponding integer to the key-worded `severity` argument.

```
1  from PIL import Image
2  import numpy as np
3  import imgaug.augmenters.imgcorruptlike as icl
4
5  image = np.asarray(Image.open("./kitty.png").convert("RGB"))
6  corruption_function = icl.ShotNoise
7  image_noise_level_01 = corruption_function(severity=1, seed=0)(image=image)
8  image_noise_level_05 = corruption_function(severity=5, seed=0)(image=image)
```

The advantage of generating data shift this way is that it can be applied to a wide range
of computer vision problems and datasets. Some of the few prerequisites for applying
this method are that the data consists of images and that these image quality corruptions
have not been used during training, for example, for data augmentation. Furthermore,
by setting the severity of the image quality corruption, we gain control over the degree
of the dataset shift. This allows us to measure how the model reacts to different degrees
of dataset shift. We can measure both how performance changes in response to dataset
shift and how calibration (introduced in *Chapter 2*) changes. We would expect that
models trained with Bayesian methods or extensions should be better calibrated, which
means that they are able to indicate to us that the data has shifted in comparison to
training and they are thus less certain in their output.

8.3.2 Revealing dataset shift with Bayesian methods

In the following code example, we will look at two of the BDL methods (Bayes by
backprop and deep ensembles) that we have encountered in the book so far and see

how they perform during the kind of artificial dataset shift described previously. We will compare their performance against a vanilla neural network.

Step 1: Preparing the environment

We start the example by importing a list of packages. This includes TensorFlow and TensorFlow Probability, which we will use for building and training the neural networks; numpy for manipulating numerical arrays (such as calculating the mean); Seaborn, Matplotlib, and pandas for plotting; cv2 and imgaug for loading and manipulating images; as well as scikit-learn for calculating the accuracy of our models.

```
1  import cv2
2  import imgaug.augmenters as iaa
3  import imgaug.augmenters.imgcorruptlike as icl
4  import matplotlib.pyplot as plt
5  import numpy as np
6  import pandas as pd
7  import seaborn as sns
8  import tensorflow as tf
9  import tensorflow_probability as tfp
10 from sklearn.metrics import accuracy_score
```

In preparation for the training, we will load the CIFAR10 dataset, which is an image classification dataset, and specify the names of the different classes. The dataset consists of 10 different classes, the names of which we specify in the following code, and provides 50,000 training images as well as 10,000 test images. We'll also save the number of training images, which will be needed to train the model with the reparameterization trick later.

```
1  cifar = tf.keras.datasets.cifar10
2  (train_images, train_labels), (test_images, test_labels) = cifar.load_data()
3
4  CLASS_NAMES = [
```

```
5        "airplane","automobile", "bird", "cat", "deer",
6        "dog", "frog", "horse", "ship", "truck"
7   ]
8
9   NUM_TRAIN_EXAMPLES = train_images.shape[0]
```

Step 2: Defining and training the models

After this prep work, we can define and train our models. We start by creating two functions to define and build the CNN. We will use these functions both for the vanilla neural network and the deep ensemble. The first function simply combines a convolutional layer with a max-pooling layer – a common approach that we introduced in *Chapter 3*.

```
1   def cnn_building_block(num_filters):
2       return tf.keras.Sequential(
3           [
4               tf.keras.layers.Conv2D(
5                   filters=num_filters, kernel_size=(3, 3), activation="relu"
6               ),
7               tf.keras.layers.MaxPool2D(strides=2),
8           ]
9       )
```

The second function then uses several convolutional/max-pooling blocks in sequence and follows this sequence with a final dense layer:

```
1   def build_and_compile_model():
2       model = tf.keras.Sequential(
3           [
4               tf.keras.layers.Rescaling(1.0 / 255, input_shape=(32, 32, 3)),
5               cnn_building_block(16),
6               cnn_building_block(32),
7               cnn_building_block(64),
```

```
 8              tf.keras.layers.MaxPool2D(strides=2),
 9              tf.keras.layers.Flatten(),
10              tf.keras.layers.Dense(64, activation="relu"),
11              tf.keras.layers.Dense(10, activation="softmax"),
12          ]
13      )
14      model.compile(
15          optimizer="adam",
16          loss="sparse_categorical_crossentropy",
17          metrics=["accuracy"],
18      )
19      return model
```

We also create two analogous functions to define and build the network using Bayes
By Backprop (BBB) based on the reparameterization trick. The strategy is the same
as for the vanilla neural network, just that we'll now use the convolutional and dense
layers from the TensorFlow Probability package instead of the TensorFlow package. The
convolutional/max-pooling blocks are then defined as follows:

```
 1  def cnn_building_block_bbb(num_filters, kl_divergence_function):
 2      return tf.keras.Sequential(
 3          [
 4              tfp.layers.Convolution2DReparameterization(
 5                  num_filters,
 6                  kernel_size=(3, 3),
 7                  kernel_divergence_fn=kl_divergence_function,
 8                  activation=tf.nn.relu,
 9              ),
10              tf.keras.layers.MaxPool2D(strides=2),
11          ]
12      )
```

And the final network is defined like this:

```
1   def build_and_compile_model_bbb():
2
3       kl_divergence_function = lambda q, p, _: tfp.distributions.kl_divergence(
4           q, p
5       ) / tf.cast(NUM_TRAIN_EXAMPLES, dtype=tf.float32)
6
7       model = tf.keras.models.Sequential(
8           [
9               tf.keras.layers.Rescaling(1.0 / 255, input_shape=(32, 32, 3)),
10              cnn_building_block_bbb(16, kl_divergence_function),
11              cnn_building_block_bbb(32, kl_divergence_function),
12              cnn_building_block_bbb(64, kl_divergence_function),
13              tf.keras.layers.Flatten(),
14              tfp.layers.DenseReparameterization(
15                  64,
16                  kernel_divergence_fn=kl_divergence_function,
17                  activation=tf.nn.relu,
18              ),
19              tfp.layers.DenseReparameterization(
20                  10,
21                  kernel_divergence_fn=kl_divergence_function,
22                  activation=tf.nn.softmax,
23              ),
24          ]
25      )
26
27      model.compile(
28          optimizer="adam",
29          loss="sparse_categorical_crossentropy",
30          metrics=["accuracy"],
31          experimental_run_tf_function=False,
```

```
32        )
33
34        model.build(input_shape=[None, 32, 32, 3])
35        return model
```

We can then train the vanilla neural network:

```
1  vanilla_model = build_and_compile_model()
2  vanilla_model.fit(train_images, train_labels, epochs=10)
```

We can also train the ensemble, with five ensemble members:

```
1  NUM_ENSEMBLE_MEMBERS = 5
2  ensemble_model = []
3  for ind in range(NUM_ENSEMBLE_MEMBERS):
4      member = build_and_compile_model()
5      print(f"Train model {ind:02}")
6      member.fit(train_images, train_labels, epochs=10)
7      ensemble_model.append(member)
```

And finally, we train the BBB model. Note that we train the BBB model for 15 instead of 10 epochs, given that it takes a little longer to converge.

```
1  bbb_model = build_and_compile_model_bbb()
2  bbb_model.fit(train_images, train_labels, epochs=15)
```

Step 3: Obtaining predictions

Now that we have three trained models, we can use them for predictions on the hold-out test set. To keep computations at a manageable degree, in this example, we will focus on the first 1,000 images in the test set:

```
1  NUM_SUBSET = 1000
2  test_images_subset = test_images[:NUM_SUBSET]
3  test_labels_subset = test_labels[:NUM_SUBSET]
```

If we want to measure the response to dataset shift, we first need to apply the artificial image corruptions to the dataset. To do that, we first specify a set of functions from the imgaug package. From their names, one can infer what type of corruption each of these functions implements: for example, the function icl.GaussianNoise corrupts an image by applying Gaussian noise to it. We also infer the number of corruption types from the number of functions and save it in the NUM_TYPES variable. Finally, we set the number of corruption levels to 5.

```
1  corruption_functions = [
2      icl.GaussianNoise,
3      icl.ShotNoise,
4      icl.ImpulseNoise,
5      icl.DefocusBlur,
6      icl.GlassBlur,
7      icl.MotionBlur,
8      icl.ZoomBlur,
9      icl.Snow,
10     icl.Frost,
11     icl.Fog,
12     icl.Brightness,
13     icl.Contrast,
14     icl.ElasticTransform,
15     icl.Pixelate,
16     icl.JpegCompression,
17  ]
18  NUM_TYPES = len(corruption_functions)
19  NUM_LEVELS = 5
```

Equipped with these functions, let us now corrupt images. In the next code block, we loop over the different corruption levels and types. We collect all corrupted images in the aptly named `corrupted_images` variable.

```
1  corrupted_images = []
2  # loop over different corruption severities
3  for corruption_severity in range(1, NUM_LEVELS+1):
4      corruption_type_batch = []
5      # loop over different corruption types
6      for corruption_type in corruption_functions:
7          corrupted_image_batch = corruption_type(
8              severity=corruption_severity, seed=0
9          )(images=test_images_subset)
10         corruption_type_batch.append(corrupted_image_batch)
11     corruption_type_batch = np.stack(corruption_type_batch, axis=0)
12     corrupted_images.append(corruption_type_batch)
13 corrupted_images = np.stack(corrupted_images, axis=0)
```

With the three models trained and the corrupted images at hand, we can now see how our models react to dataset shift of different levels. We will first obtain predictions on the corrupted images from the three models. To run inference, we need to reshape the corrupted images to an input shape that is accepted by the models for inference. At the moment, the images are still stored on different axes for the corruption types and levels. We change this by reshaping the `corrupted_images` array:

```
1  corrupted_images = corrupted_images.reshape((-1, 32, 32, 3))
```

Then we can perform inference with the vanilla CNN model, both on the original images and the corrupted images. After we have inferred the model predictions, we reshape the predictions in order to separate predictions for the corruption types and levels:

```
1  # Get predictions on original images
2  vanilla_predictions = vanilla_model.predict(test_images_subset)
3  # Get predictions on corrupted images
4  vanilla_predictions_on_corrupted = vanilla_model.predict(corrupted_images)
5  vanilla_predictions_on_corrupted = vanilla_predictions_on_corrupted.reshape(
6      (NUM_LEVELS, NUM_TYPES, NUM_SUBSET, -1)
7  )
```

To run inference with the ensemble model, we first define a prediction function to avoid code duplication. This function handles the looping over the different member models of the ensemble and combines the different predictions in the end via averaging:

```
1  def get_ensemble_predictions(images, num_inferences):
2      ensemble_predictions = tf.stack(
3          [
4              ensemble_model[ensemble_ind].predict(images)
5              for ensemble_ind in range(num_inferences)
6          ],
7          axis=0,
8      )
9      return np.mean(ensemble_predictions, axis=0)
```

Equipped with this function, we can perform inference with the ensemble model on both the original and corrupted images:

```
1  # Get predictions on original images
2  ensemble_predictions = get_ensemble_predictions(
3      test_images_subset, NUM_ENSEMBLE_MEMBERS
4  )
5  # Get predictions on corrupted images
6  ensemble_predictions_on_corrupted = get_ensemble_predictions(
7      corrupted_images, NUM_ENSEMBLE_MEMBERS
```

```
 8  )
 9  ensemble_predictions_on_corrupted = ensemble_predictions_on_corrupted.reshape(
10      (NUM_LEVELS, NUM_TYPES, NUM_SUBSET, -1)
11  )
```

Just as for the ensemble model, we write an inference function for the BBB model, which handles the iteration over different sampling loops and collects and combines the results:

```
1  def get_bbb_predictions(images, num_inferences):
2      bbb_predictions = tf.stack(
3          [bbb_model.predict(images) for _ in range(num_inferences)],
4          axis=0,
5      )
6      return np.mean(bbb_predictions, axis=0)
```

We then put this function to use to obtain the BBB model predictions on the original and corrupted images. We sample from the BBB model 20 times:

```
 1  NUM_INFERENCES_BBB = 20
 2  # Got predictions on original images
 3  bbb_predictions = get_bbb_predictions(
 4      test_images_subset, NUM_INFERENCES_BBB
 5  )
 6  # Get predictions on corrupted images
 7  bbb_predictions_on_corrupted = get_bbb_predictions(
 8      corrupted_images, NUM_INFERENCES_BBB
 9  )
10  bbb_predictions_on_corrupted = bbb_predictions_on_corrupted.reshape(
11      (NUM_LEVELS, NUM_TYPES, NUM_SUBSET, -1)
12  )
```

We can convert the predictions of the three models to predicted classes and associated confidence scores by returning the index of the class with the maximum softmax score and the maximum softmax score, respectively:

```
1  def get_classes_and_scores(model_predictions):
2      model_predicted_classes = np.argmax(model_predictions, axis=-1)
3      model_scores = np.max(model_predictions, axis=-1)
4      return model_predicted_classes, model_scores
```

This function can then be applied to get the predicted classes and confidence scores for our three models:

```
1  # Vanilla model
2  vanilla_predicted_classes, vanilla_scores = get_classes_and_scores(
3      vanilla_predictions
4  )
5  (
6      vanilla_predicted_classes_on_corrupted,
7      vanilla_scores_on_corrupted,
8  ) = get_classes_and_scores(vanilla_predictions_on_corrupted)
9
10  # Ensemble model
11  (
12      ensemble_predicted_classes,
13      ensemble_scores,
14  ) = get_classes_and_scores(ensemble_predictions)
15  (
16      ensemble_predicted_classes_on_corrupted,
17      ensemble_scores_on_corrupted,
18  ) = get_classes_and_scores(ensemble_predictions_on_corrupted)
19
20  # BBB model
```

```
21  (
22      bbb_predicted_classes,
23      bbb_scores,
24  ) = get_classes_and_scores(bbb_predictions)
25  (
26      bbb_predicted_classes_on_corrupted,
27      bbb_scores_on_corrupted,
28  ) = get_classes_and_scores(bbb_predictions_on_corrupted)
```

Let us visualize what these predicted classes and confidence scores look like for the three models on a selected image showing an automobile. For plotting, we first reshape the array that contains the corrupted images to a more convenient format:

```
1  plot_images = corrupted_images.reshape(
2      (NUM_LEVELS, NUM_TYPES, NUM_SUBSET, 32, 32, 3)
3  )
```

We then plot the selected automobile image with the first three corruption types in the list across all five corruption levels. For each combination, we display in the image title the predicted score of each model and in squared parentheses the predicted class. The plot is shown in *Figure 8.5*.

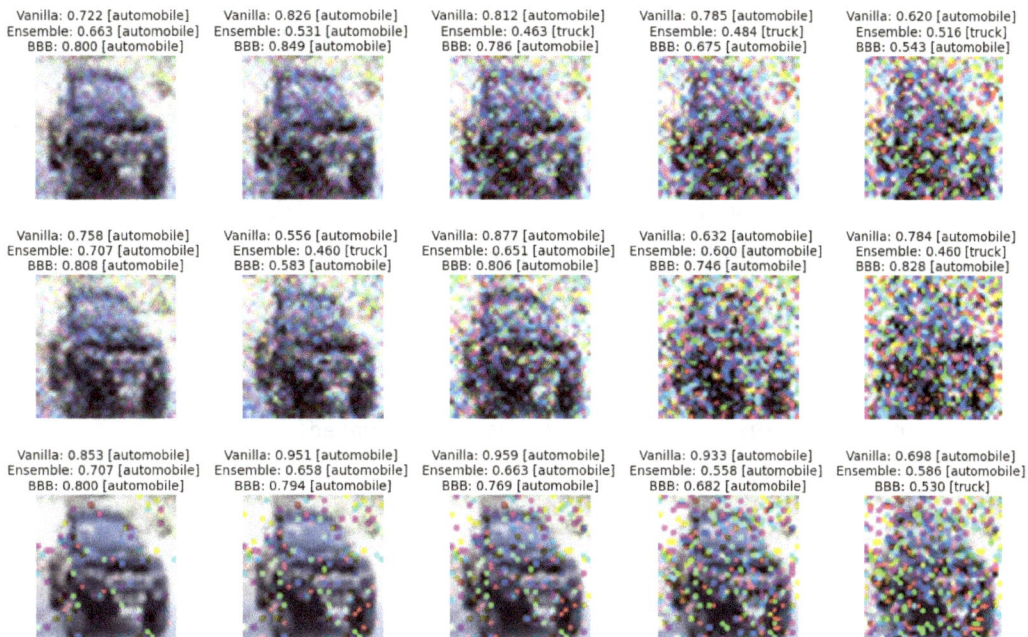

Figure 8.5: An automobile image has been corrupted with different corruption types (rows) and levels (columns, severity increases from left to right)

The code continues:

```
1  # Index of the selected images
2  ind_image = 9
3  # Define figure
4  fig, axes = plt.subplots(nrows=3, ncols=5, figsize=(16, 10))
5  # Loop over corruption levels
6  for ind_level in range(NUM_LEVELS):
7      # Loop over corruption types
8      for ind_type in range(3):
9          # Plot slightly upscaled image for easier inspection
10         image = plot_images[ind_level, ind_type, ind_image, ...]
11         image_upscaled = cv2.resize(
12             image, dsize=(150, 150), interpolation=cv2.INTER_CUBIC
```

```
13          )
14          axes[ind_type, ind_level].imshow(image_upscaled)
15          # Get score and class predicted by vanilla model
16          vanilla_score = vanilla_scores_on_corrupted[
17              ind_level, ind_type, ind_image, ...
18          ]
19          vanilla_prediction = vanilla_predicted_classes_on_corrupted[
20              ind_level, ind_type, ind_image, ...
21          ]
22          # Get score and class predicted by ensemble model
23          ensemble_score = ensemble_scores_on_corrupted[
24              ind_level, ind_type, ind_image, ...
25          ]
26          ensemble_prediction = ensemble_predicted_classes_on_corrupted[
27              ind_level, ind_type, ind_image, ...
28          ]
29          # Get score and class predicted by BBB model
30          bbb_score = bbb_scores_on_corrupted[ind_level, ind_type, ind_image, ...]
31          bbb_prediction = bbb_predicted_classes_on_corrupted[
32              ind_level, ind_type, ind_image, ...
33          ]
34          # Plot prediction info in title
35          title_text = (
36              f"Vanilla: {vanilla_score:.3f} "
37              + f"[{CLASS_NAMES[vanilla_prediction]}] \n"
38              + f"Ensemble: {ensemble_score:.3f} "
39              + f"[{CLASS_NAMES[ensemble_prediction]}] \n"
40              + f"BBB: {bbb_score:.3f} "
41              + f"[{CLASS_NAMES[bbb_prediction]}]"
42          )
43          axes[ind_type, ind_level].set_title(title_text, fontsize=14)
44          # Remove axes ticks and labels
45          axes[ind_type, ind_level].axis("off")
```

```
46  fig.tight_layout()
47  plt.show()
```

Figure 8.5 only shows results for a single image, so we should not read too much into these results. However, we can already observe that the prediction scores for the two Bayesian methods (and especially the ensemble method) tend to be less extreme than for the vanilla neural network, which has predicted scores as high as 0.95. Furthermore, we see that, for all three models, prediction scores usually decrease as the corruption level increases. This is expected: given that the car in the image becomes less discernible with more corruption, we would want the model to become less confident as well. In particular, the ensemble method shows a nice and consistent decrease in predicted scores with increased corruption levels.

Step 4: Measuring accuracy

Are some models more robust to dataset shift than other models? We can answer this question by looking at the accuracy of the three models at different corruptions levels. It is expected that all models will show lower accuracy as the input image becomes more and more corrupted. However, more robust models should lose less in accuracy as the corruptions become more severe.

First, we can calculate the accuracy of the three models on the original test images:

```
1  vanilla_acc = accuracy_score(
2      test_labels_subset.flatten(), vanilla_predicted_classes
3  )
4  ensemble_acc = accuracy_score(
5      test_labels_subset.flatten(), ensemble_predicted_classes
6  )
7  bbb_acc = accuracy_score(
8      test_labels_subset.flatten(), bbb_predicted_classes
9  )
```

We can store these accuracies in a list of dictionaries, which will make it easier to plot them systematically. We pass the respective name of the models. For corruption type and level, we pass 0 because these are the accuracies on the original images.

```
1  accuracies = [
2      {"model_name": "vanilla", "type": 0, "level": 0, "accuracy": vanilla_acc},
3      {"model_name": "ensemble", "type": 0, "level": 0, "accuracy": ensemble_acc},
4      {"model_name": "bbb", "type": 0, "level": 0, "accuracy": bbb_acc},
5  ]
```

Next, we calculate the accuracy of the three models on the different corruption type by corruption level combinations. We also append the results to the list of accuracies that we started previously:

```
1  for ind_type in range(NUM_TYPES):
2      for ind_level in range(NUM_LEVELS):
3          # Calculate accuracy for vanilla model
4          vanilla_acc_on_corrupted = accuracy_score(
5              test_labels_subset.flatten(),
6              vanilla_predicted_classes_on_corrupted[ind_level, ind_type, :],
7          )
8          accuracies.append(
9              {
10                 "model_name": "vanilla",
11                 "type": ind_type + 1,
12                 "level": ind_level + 1,
13                 "accuracy": vanilla_acc_on_corrupted,
14             }
15         )
16
17         # Calculate accuracy for ensemble model
18         ensemble_acc_on_corrupted = accuracy_score(
```

```
19              test_labels_subset.flatten(),
20              ensemble_predicted_classes_on_corrupted[ind_level, ind_type, :],
21          )
22          accuracies.append(
23              {
24                  "model_name": "ensemble",
25                  "type": ind_type + 1,
26                  "level": ind_level + 1,
27                  "accuracy": ensemble_acc_on_corrupted,
28              }
29          )
30
31          # Calculate accuracy for BBB model
32          bbb_acc_on_corrupted = accuracy_score(
33              test_labels_subset.flatten(),
34              bbb_predicted_classes_on_corrupted[ind_level, ind_type, :],
35          )
36          accuracies.append(
37              {
38                  "model_name": "bbb",
39                  "type": ind_type + 1,
40                  "level": ind_level + 1,
41                  "accuracy": bbb_acc_on_corrupted,
42              }
43          )
```

We can then plot the distributions of accuracies for the original images and the increasingly corrupted images. We first convert the list of dictionaries to a pandas dataframe. This has the advantage that the dataframe can be directly passed to the plotting package seaborn. This allows us to specify that we want to plot the different models' results in different hues.

```
1  df = pd.DataFrame(accuracies)
2  plt.figure(dpi=100)
3  sns.boxplot(data=df, x="level", y="accuracy", hue="model_name")
4  plt.legend(loc="center left", bbox_to_anchor=(1, 0.5))
5  plt.tight_layout
6  plt.show()
```

This produces the following output:

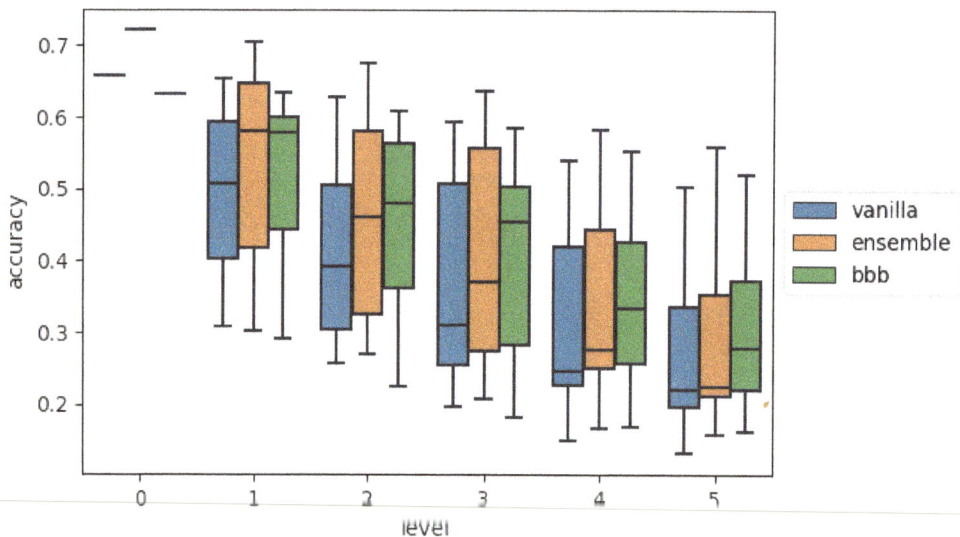

Figure 8.6: Accuracy for the three different models (different hues) for the original test images (level 0) as well as for increasing levels of corruption (level 1-5)

The resulting plot is shown in *Figure 8.6*. We can see that, on the original test images, the vanilla and BBB model have comparable accuracy, while the ensemble model has slightly higher accuracy. As corruption is introduced, we see that the performance of the vanilla neural network is worse (often significantly) than the performance of the ensemble or BBB. This relative improvement in performance of the BDL models demonstrates the regularization effect of Bayesian methods: these methods are able to capture the distribution of the data more effectively, making them more robust to perturbations. BBB exhibits particular resilience to increasing amounts of data corruption, demonstrating a key benefit of variational learning.

Step 5: Measuring calibration

Looking at accuracy is a good way to determine how robust a model is against dataset shift. But it does not really tell us whether the models are capable of signalling to us (via lower confidence scores) when the dataset has shifted and the models have become less confident in their output. This question can be answered by looking at how well models remain calibrated under dataset shift. We introduced calibration and expected calibration errors on a conceptual level back in *Chapter 3*. We are now going to put these concepts into practice to understand whether models adjust their confidence appropriately as the images become increasingly corrupted and hard to predict.

First, we will implement the Expected Calibration Error (ECE) introduced in *Chapter 3* as a scalar measure of calibration:

```
1  def expected_calibration_error(
2      pred_correct,
3      pred_score,
4      n_bins=5,
5  ):
6      """Compute expected calibration error.
7      ----------
8      pred_correct : np.ndarray (n_samples,)
9          Whether the prediction is correct or not
10     pred_score : np.ndarray (n_samples,)
11         Confidence in the prediction
12     n_bins : int, default=5
13         Number of bins to discretize the [0, 1] interval.
14     """
15     # Convert from bool to integer (makes counting easier)
16     pred_correct = pred_correct.astype(np.int32)
```

```
17
18     # Create bins and assign prediction scores to bins
19     bins = np.linspace(0.0, 1.0, n_bins + 1)
20     binids = np.searchsorted(bins[1:-1], pred_score)
21
22     # Count number of samples and correct predictions per bin
23     bin_true_counts = np.bincount(
24         binids, weights=pred_correct, minlength=len(bins)
25     )
26     bin_counts = np.bincount(binids, minlength=len(bins))
27
28     # Calculate sum of confidence scores per bin
29     bin_probs = np.bincount(binids, weights=pred_score, minlength=len(bins))
30
31     # Identify bins that contain samples
32     nonzero = bin_counts != 0
33     # Calculate accuracy for every bin
34     bin_acc = bin_true_counts[nonzero] / bin_counts[nonzero]
35     # Calculate average confidence scores per bin
36     bin_conf = bin_probs[nonzero] / bin_counts[nonzero]
37
38     return np.average(np.abs(bin_acc - bin_conf), weights=bin_counts[nonzero])
```

We can then calculate ECE for the three models on the original test images. We set the number of bins to 10, which is a common choice for calculating ECE:

```
1  NUM_BINS = 10
2
3  vanilla_cal = expected_calibration_error(
4      test_labels_subset.flatten() == vanilla_predicted_classes,
5      vanilla_scores,
6      n_bins=NUM_BINS,
7  )
```

```
 8
 9  ensemble_cal = expected_calibration_error(
10      test_labels_subset.flatten() == ensemble_predicted_classes,
11      ensemble_scores,
12      n_bins=NUM_BINS,
13  )
14
15  bbb_cal = expected_calibration_error(
16      test_labels_subset.flatten() == bbb_predicted_classes,
17      bbb_scores,
18      n_bins=NUM_BINS,
19  )
```

Just as we did for the accuracies earlier, we will store the calibration results in a list of dictionaries, which will make it easier to plot them:

```
 1  calibration = [
 2      {
 3          "model_name": "vanilla",
 4          "type": 0,
 5          "level": 0,
 6          "calibration_error": vanilla_cal,
 7      },
 8      {
 9          "model_name": "ensemble",
10          "type": 0,
11          "level": 0,
12          "calibration_error": ensemble_cal,
13      },
14      {
15          "model_name": "bbb",
16          "type": 0,
17          "level": 0,
```

```
18            "calibration_error": bbb_cal,
19        },
20    ]
```

Next, we calculate the expected calibration error of the three models on the different corruption types by corruption level combinations. We also append the results to the list of calibration results that we started previously:

```
1   for ind_type in range(NUM_TYPES):
2       for ind_level in range(NUM_LEVELS):
3           # Calculate calibration error for vanilla model
4           vanilla_cal_on_corrupted = expected_calibration_error(
5               test_labels_subset.flatten()
6               == vanilla_predicted_classes_on_corrupted[ind_level, ind_type, :],
7               vanilla_scores_on_corrupted[ind_level, ind_type, :],
8           )
9           calibration.append(
10              {
11                  "model_name": "vanilla",
12                  "type": ind_type + 1,
13                  "level": ind_level + 1,
14                  "calibration_error": vanilla_cal_on_corrupted,
15              }
16          )
17
18          # Calculate calibration error for ensemble model
19          ensemble_cal_on_corrupted = expected_calibration_error(
20              test_labels_subset.flatten()
21              == ensemble_predicted_classes_on_corrupted[ind_level, ind_type, :],
22              ensemble_scores_on_corrupted[ind_level, ind_type, :],
23          )
24          calibration.append(
25              {
```

```
26                "model_name": "ensemble",
27                "type": ind_type + 1,
28                "level": ind_level + 1,
29                "calibration_error": ensemble_cal_on_corrupted,
30            }
31        )
32
33        # Calculate calibration error for BBB model
34        bbb_cal_on_corrupted = expected_calibration_error(
35            test_labels_subset.flatten()
36            == bbb_predicted_classes_on_corrupted[ind_level, ind_type, :],
37            bbb_scores_on_corrupted[ind_level, ind_type, :],
38        )
39        calibration.append(
40            {
41                "model_name": "bbb",
42                "type": ind_type + 1,
43                "level": ind_level + 1,
44                "calibration_error": bbb_cal_on_corrupted,
45            }
46        )
```

Finally, we plot the calibration results in a boxplot, again using pandas and seaborn:

```
1  df = pd.DataFrame(calibration)
2  plt.figure(dpi=100)
3  sns.boxplot(data=df, x="level", y="calibration_error", hue="model_name")
4  plt.legend(loc="center left", bbox_to_anchor=(1, 0.5))
5  plt.tight_layout
6  plt.show()
```

The calibration results are shown in *Figure 8.7*. We can see that, on the original test images, all three models have relatively low calibration error, with the ensemble model performing slightly worse than the two other models. As we apply increasing levels of dataset shift, we can see that calibration error increases by a lot for the vanilla model. For the two Bayesian methods, calibration error also increases but by much less than for the vanilla model. This means that the Bayesian methods are better at indicating (via lower confidence scores) when the dataset has shifted and that the Bayesian models become relatively less confident in their output with increased corruption (as they should).

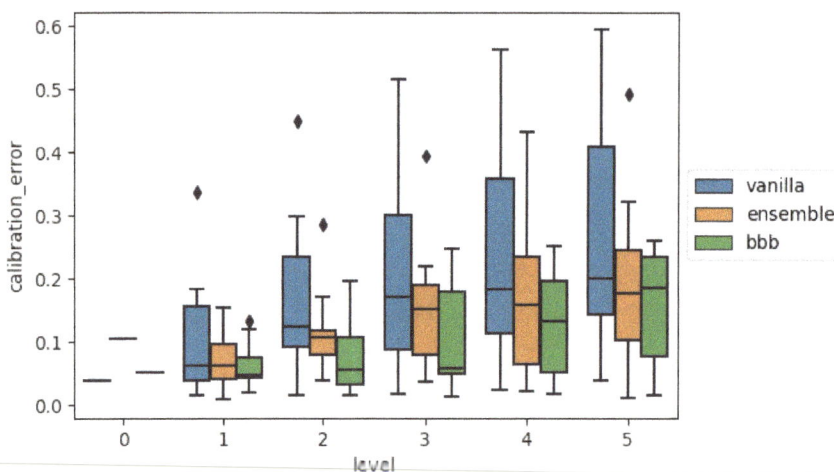

Figure 8.7: Expected calibration error for the three different models for the original test images (level 0) as well as for increasing levels of corruption (level 1-5)

In the next section, we will look into data selection.

8.4 Using data selection via uncertainty to keep models fresh

We saw at the beginning of the chapter that we can use uncertainties to figure out whether data is part of the training data or not. We can expand on this idea in the context of an area of machine learning called **active learning**. The promise of active learning is that a model can learn more effectively on less data if we have a way to control the type of data it is trained on. Conceptually, this makes sense: if we train a model on data that is not of sufficient quality, it will also not perform well. Active learning is a way to guide the learning process and data a model is trained on by providing functions that can acquire data from a pool of data that is not part of the training data. By iteratively selecting the right data from the pool, we can train a model that performs better than if we had chosen the data from the pool at random.

Active learning can be used in many modern-day systems where there is a ton of unlabeled data available and we need to carefully select the amount of data we want to label. An example is an autonomous driving system: the camera on the car records a lot of data, but there is typically no budget to label all of it. By carefully choosing the most informative data points, we can improve the model performance at a lower cost than when we would have randomly selected the data to label. In the context of active learning, estimating uncertainties plays an important role. A model will typically learn more from areas of the data distribution that were predicted with low confidence. Let's look at a case study to see how we can use uncertainty in the context of active learning.

In this case study, we will reproduce the results from a fundamental active learning paper: *Deep Bayesian Active Learning with Image Data* (2017). We will use the MNIST dataset and train a model on more and more data, where we select the data points to add to our training set via an uncertainty method. In this case, we will use epistemic uncertainty to select the most informative data points. Images with high epistemic uncertainty should be images that the model did not see before; the uncertainty can be reduced by adding more of them. As a comparison, we will also select data points at random.

Step 1: Preparing our dataset

We will start by creating our functions to load the dataset. The dataset functions need the following library imports:

```
1  import dataclasses
2  from pathlib import Path
3  import uuid
4  from typing import Optional, Tuple
5
6  import numpy as np
7  import tensorflow as tf
8  from sklearn.utils import shuffle
```

As our total dataset will have quite a few components, we will create a small dataclass to easily access all the different parts of our dataset. We will also modify the __repr__ function of the dataclass. This allows us to print the content of the dataset in a more readable format.

```
1  @dataclasses.dataclass
2  class Data:
3      x_train: np.ndarray
4      y_train: np.ndarray
5      x_test: np.ndarray
6      y_test: np.ndarray
7      x_train_al: Optional[np.ndarray] = None
8      y_train_al: Optional[np.ndarray] = None
9
10     def __repr__(self) -> str:
11         repr_str = ""
12         for field in dataclasses.fields(self):
13             repr_str += f"{field.name}: {getattr(self, field.name).shape} \n"
14         return repr_str
```

We can then define our function to load our standard dataset.

```python
def get_data() -> Data:
    num_classes = 10
    (x_train, y_train), (x_test, y_test) = tf.keras.datasets.mnist.load_data()
    # Scale images to the [0, 1] range
    x_train = x_train.astype("float32") / 255
    x_test = x_test.astype("float32") / 255
    # Make sure images have shape (28, 28, 1)
    x_train = np.expand_dims(x_train, -1)
    x_test = np.expand_dims(x_test, -1)
    y_train = tf.keras.utils.to_categorical(y_train, num_classes)
    y_test = tf.keras.utils.to_categorical(y_test, num_classes)
    return Data(x_train, y_train, x_test, y_test)
```

Initially, we will start training on just 20 samples from the *MNIST* dataset. We will then acquire 10 data points at a time, and retrain our model again. To help our model a little bit in the beginning, we will make sure that the 20 data points are balanced across the different classes of the dataset. The following function gives us the indices that we can use to create the initial 20 samples, 2 samples of each class:

```python
def get_random_balanced_indices(
    data: Data, initial_n_samples: int
) -> np.ndarray:
    labels = np.argmax(data.y_train, axis=1)
    indices = []
    label_list = np.unique(labels)
    for label in label_list:
        indices_label = np.random.choice(
            np.argwhere(labels == label).flatten(),
            size=initial_n_samples // len(label_list),
            replace=False
```

```
12              )
13              indices.extend(indices_label)
14          indices = np.array(indices)
15          np.random.shuffle(indices)
16          return indices
```

We can then define a small function to actually get our initial dataset:

```
1  def get_initial_ds(data: Data, initial_n_samples: int) -> Data:
2      indices = get_random_balanced_indices(data, initial_n_samples)
3      x_train_al, y_train_al = data.x_train[indices], data.y_train[indices]
4      x_train = np.delete(data.x_train, indices, axis=0)
5      y_train = np.delete(data.y_train, indices, axis=0)
6      return Data(
7          x_train, y_train, data.x_test, data.y_test, x_train_al, y_train_al
8      )
```

Step 2: Setting up our configuration

Before we start to build our model and create the active learning loop, we define a small configuration dataclass to store some main variables we might want to play around with when running our active learning script. Creating configuration classes such as these allows you to play around with different parameters.

```
1  @dataclasses.dataclass
2  class Config:
3      initial_n_samples: int
4      n_total_samples: int
5      n_epochs: int
6      n_samples_per_iter: int
7      # string representation of the acquisition function
8      acquisition_type: str
9      # number of mc_dropout iterations
```

```
10        n_iter: int
```

Step 3: Defining the model

We can now define our model. We will use a small, simple CNN with dropout.

```
1  def build_model():
2      model = tf.keras.models.Sequential([
3          Input(shape=(28, 28, 1)),
4          layers.Conv2D(32, kernel_size=(4, 4), activation="relu"),
5          layers.Conv2D(32, kernel_size=(4, 4), activation="relu"),
6          layers.MaxPooling2D(pool_size=(2, 2)),
7          layers.Dropout(0.25),
8          layers.Flatten(),
9          layers.Dense(128, activation="relu"),
10         layers.Dropout(0.5),
11         layers.Dense(10, activation="softmax"),
12     ])
13     model.compile(
14        tf.keras.optimizers.Adam(),
15        loss="categorical_crossentropy",
16        metrics=["accuracy"],
17        experimental_run_tf_function=False,
18     )
19     return model
```

Step 4: Defining the uncertainty functions

As indicated, we will use epistemic uncertainty (also knowledge uncertainty) as our main uncertainty function to acquire new samples. Let's define the function to compute epistemic uncertainty over our predictions. We assume that the input predictions (preds) are of shape n_images, n_predictions, n_classes. We first define a function to compute total uncertainty. Given an ensemble of model predictions, this can be defined as the entropy of the averaged predictions of the ensemble.

```
1  def total_uncertainty(
2      preds: np.ndarray, epsilon: float = 1e-10
3  ) -> np.ndarray:
4      mean_preds = np.mean(preds, axis=1)
5      log_preds = -np.log(mean_preds + epsilon)
6      return np.sum(mean_preds * log_preds, axis=1)
```

We then define data uncertainty (or aleatoric uncertainty), which for an ensemble is the average of the entropy of each ensemble member.

```
1  def data_uncertainty(preds: np.ndarray, epsilon: float = 1e-10) -> np.ndarray:
2      log_preds = -np.log(preds + epsilon)
3      return np.mean(np.sum(preds * log_preds, axis=2), axis=1)
```

Finally, we have our knowledge (or epistemic) uncertainty, which is simply subtracting data uncertainty from the total uncertainty of the predictions.

```
1  def knowledge_uncertainty(
2      preds: np.ndarray, epsilon: float = 1e-10
3  ) -> np.ndarray:
4      return total_uncertainty(preds, epsilon) - data_uncertainty(preds, epsilon)
```

With these uncertainty functions defined, we can define the actual acquisition functions that take as main input our training data and our model. To acquire samples via knowledge uncertainty, we do the following:

1. Obtain our ensemble of predictions via MC dropout.

2. Compute the knowledge uncertainty values over this ensemble.

3. Sort the uncertainty values, get their index and return the indices of our training data with the highest epistemic uncertainty.

We can then, later on, reuse these indices to index into our training data and actually acquire the training samples we want to add.

```python
from typing import Callable
from keras import Model
from tqdm import tqdm

import numpy as np

def acquire_knowledge_uncertainty(
    x_train: np.ndarray,
    n_samples: int,
    model: Model,
    n_iter: int,
    *args,
    **kwargs
):
    preds = get_mc_predictions(model, n_iter, x_train)
    ku = knowledge_uncertainty(preds)
    return np.argsort(ku, axis=-1)[-n_samples:]
```

We obtain our MC dropout predictions as follows:

```python
def get_mc_predictions(
    model: Model, n_iter: int, x_train: np.ndarray
) -> np.ndarray:
    preds = []
    for _ in tqdm(range(n_iter)):
        preds_iter = [
            model(batch, training=True)
            for batch in np.array_split(x_train, 6)
        ]
        preds.append(np.concatenate(preds_iter))
```

```
11      # format data such that we have n_images, n_predictions, n_classes
12      preds = np.moveaxis(np.stack(preds), 0, 1)
13      return preds
```

To avoid running out of memory, we iterate over our training data in batches of six, where for every batch we compute our predictions n_iter times. To make sure that our predictions are varied, we set the model's training parameter to True.

For our comparison, we define an acquisition function that returns a random number of indices as well:

```
1   def acquire_random(x_train: np.ndarray, n_samples: int, *args, **kwargs):
2       return np.random.randint(low=0, high=len(x_train), size=n_samples)
```

Finally, we define a small function according to the *factory method pattern* to make sure that we can use the same function in our loop to use either the random acquisition function or knowledge uncertainty. Small factory functions such as these help to keep your code modular when you want to run the same code with different configurations.

```
1   def acquisition_factory(acquisition_type: str) -> Callable:
2       if acquisition_type == "knowledge_uncertainty":
3           return acquire_knowledge_uncertainty
4       if acquisition_type == "random":
5           return acquire_random
```

Now that we have defined our acquisition functions, we are ready to actually define the loop that runs our active learning iterations.

Step 5: Defining the loop

Let's start by defining our configuration. In this case, we are using knowledge uncertainty as our uncertainty function. In a different loop, we will use a random acquisition function to compare the results of the loop we are about to define. We will start our dataset with

20 samples until we reach a total of 1,000 samples. Each model will be trained for 50 epochs and per iteration, we acquire 10 samples. To obtain our MC dropout predictions, we will run over our full training set (minus the already acquired samples) 100 times.

```
1  cfg = Config(
2      initial_n_samples=20,
3      n_total_samples=1000,
4      n_epochs=50,
5      n_samples_per_iteration=10,
6      acquisition_type="knowledge_uncertainty",
7      n_iter=100,
8  )
```

We can then get our data and define an empty dictionary to keep track of the test accuracy per iteration. We also create an empty list to keep track of the full list of indices we added to our training data.

```
1  data: Data = get_initial_ds(get_data(), cfg.initial_n_samples)
2  accuracies = {}
3  added_indices = []
```

We also assign a **universally unique identifier (UUID)** to our run to make sure we can discover it easily and do not overwrite the outcomes we save as part of our loop. We create the directory where we will save our data and save our configuration in that directory to ensure that we always know with what kind of configuration the data in our model_dir was created.

```
1  run_uuid = str(uuid.uuid4())
2  model_dir = Path("./models") / cfg.acquisition_type / run_uuid
3  model_dir.mkdir(parents=True, exist_ok=True)
```

We can now actually run our active learning loop. We will break this loop into three sections:

1. We define the loop and fit a model on the acquired samples:

```
1   for i in range(cfg.n_total_samples // cfg.n_samples_per_iter):
2       iter_dir = model_dir / str(i)
3       model = build_model()
4       model.fit(
5           x=data.x_train_al,
6           y=data.y_train_al,
7           validation_data=(data.x_test, data.y_test),
8           epochs=cfg.n_epochs,
9           callbacks=[get_callback(iter_dir)],
10          verbose=2,
11      )
```

2. We then load the model with the best validation accuracy and update our dataset based on the acquisition function:

```
1       model = tf.keras.models.load_model(iter_dir)
2       indices_to_add = acquisition_factory(cfg.acquisition_type)(
3           data.x_train,
4           cfg.n_samples_per_iter,
5           n_iter=cfg.n_iter,
6           model=model,
7       )
8       added_indices.append(indices_to_add)
9       data, (iter_x, iter_y) = update_ds(data, indices_to_add)
```

3. We finally save the added images, compute the test accuracy, and save the results:

```
1    save_images_and_labels_added(iter_dir, iter_x, iter_y)
2    preds = model(data.x_test)
3    accuracy = get_accuracy(data.y_test, preds)
4    accuracies[i] = accuracy
5    save_results(accuracies, added_indices, model_dir)
```

In this loop, we defined a few small helper functions. First of all, we defined a callback for our model to save the model with the highest validation accuracy to our model directory:

```
1    def get_callback(model_dir: Path):
2        model_checkpoint_callback = tf.keras.callbacks.ModelCheckpoint(
3            str(model_dir),
4            monitor="val_accuracy",
5            verbose=0,
6            save_best_only=True,
7        )
8        return model_checkpoint_callback
```

We also defined a function to compute the accuracy of our test set:

```
1    def get_accuracy(y_test: np.ndarray, preds: np.ndarray) -> float:
2        acc = tf.keras.metrics.CategoricalAccuracy()
3        acc.update_state(preds, y_test)
4        return acc.result().numpy() * 100
```

And we defined two small functions to save the results per iteration:

```
1  def save_images_and_labels_added(
2      output_path: Path, iter_x: np.ndarray, iter_y: np.ndarray
3  ):
4      df = pd.DataFrame()
5      df["label"] = np.argmax(iter_y, axis=1)
6      iter_x_normalised = (np.squeeze(iter_x, axis=-1) * 255).astype(np.uint8)
7      df["image"] = iter_x_normalised.reshape(10, 28*28).tolist()
8      df.to_parquet(output_path / "added.parquet", index=False)
9
10 def save_results(
11     accuracies: Dict[int, float], added_indices: List[int], model_dir: Path
12 ):
13     df = pd.DataFrame(accuracies.items(), columns=["i", "accuracy"])
14     df["added"] = added_indices
15     df.to_parquet(f"{model_dir}/results.parquet", index=False)
```

Note that running the active learning loop takes quite a long time: for every iteration, we train and evaluate our model for 50 epochs, and then run through our pool set (the full training dataset minus the acquired samples) 100 times. When using a random acquisition function, we avoid the last step but still run our validation data through our model 50 times per iteration, just to make sure that we use the model with the best validation accuracy. This takes time, but picking the model with just the best *training* accuracy would be risky: our model sees the same few images many times during training and is therefore likely to overfit to the training data.

Step 6: Inspecting the results

Now that we have our loop, we can inspect the results of this process. We will use seaborn and matplotlib to visualize our results:

```
1  import seaborn as sns
2  import matplotlib.pyplot as plt
3  import pandas as pd
4  import numpy as np
5  sns.set_style("darkgrid")
6  sns.set_context("paper")
```

The main result we are interested in is the test accuracy over time for both the models trained with a random acquisition function and the models trained with data acquired via knowledge uncertainty. To visualize this, we define a function that loads the results and then returns a plot that shows the accuracy per active learning iteration cycle:

```
1  def plot(uuid: str, acquisition: str, ax=None):
2      acq_name = acquisition.replace("_", " ")
3      df = pd.read_parquet(f"./models/{acquisition}/{uuid}/results.parquet")[:-1]
4      df = df.rename(columns={"accuracy": acq_name})
5      df["n_samples"] = df["i"].apply(lambda x: x*10 + 20)
6      return df.plot.line(
7          x="n_samples", y=acq_name, style='.-', figsize=(8,5), ax=ax
8      )
```

We can then use this function to plot the results for both acquisition functions:

```
1  ax = plot("bc1adec5-bc34-44a6-a0eb-fa7cb67854e4", "random")
2  ax = plot(
3      "5c8d6001-a5fb-45d3-a7cb-2a8a46b93d18", "knowledge_uncertainty", ax=ax
4  )
5  plt.xticks(np.arange(0, 1050, 50))
6  plt.yticks(np.arange(54, 102, 2))
7  plt.ylabel("Accuracy")
8  plt.xlabel("Number of acquired samples")
```

```
9  plt.show()
```

This produces the following output:

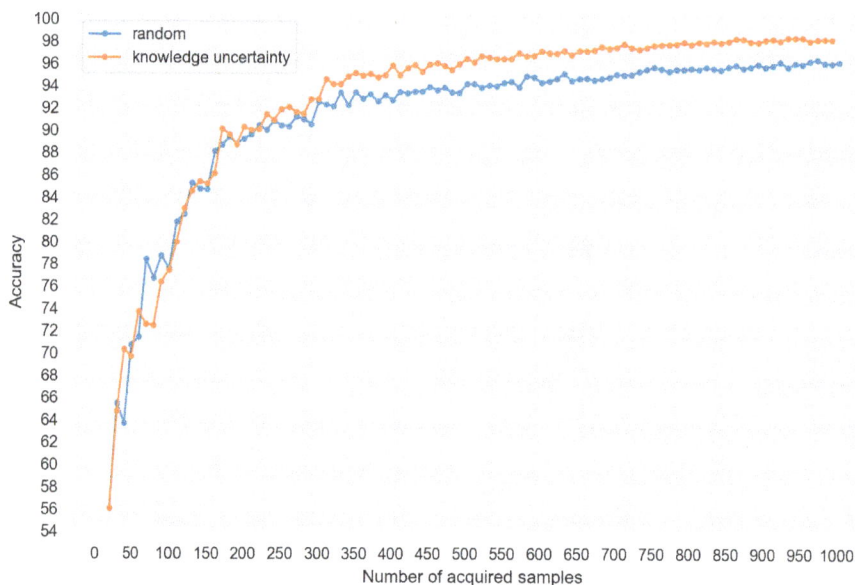

Figure 8.8: Active learning results

Figure 8.8 shows that acquiring samples via knowledge uncertainty starts to improve the model's accuracy significantly after around 300 acquired samples. The final accuracy of this model is about two percentage points higher than the accuracy of the model trained on random samples. This might not look like a lot, but we can also look at the data in another way: how many samples were needed to achieve a particular accuracy? If we inspect the plot, we can see that the knowledge uncertainty line achieves an accuracy of 96% with 400 training samples. The model trained on random samples required at least 750 samples to achieve the same accuracy. That's almost double the amount of data for the same accuracy. This shows that active learning with the right acquisition function can be very useful, specifically in cases where compute resources are available, but labeling is expensive: with the right samples, we might be able to decrease our labeling cost by a factor of two to achieve the same accuracy.

Because we saved the acquired samples for every iteration, we can also inspect the type of images selected by both models. To make our visualization easier to interpret, we will visualize the last five acquired images for every method for every label. To do this, we first define a function that returns the images per label for a set of model directories:

```python
def get_imgs_per_label(model_dirs) -> Dict[int, np.ndarray]:
    imgs_per_label = {i: [] for i in range(10)}
    for model_dir in model_dirs:
        df = pd.read_parquet(model_dir / "images_added.parquet")
        df.image = df.image.apply(
            lambda x: x.reshape(28, 28).astype(np.uint8)
        )
        for label in df.label.unique():
            dff = df[df.label == label]
            if len(dff) == 0:
                continue
            imgs_per_label[label].append(np.hstack(dff.image))
    return imgs_per_label
```

We then define a function that creates a `PIL Image` where we concatenate the images per label for a particular acquisition function:

```python
from PIL import Image
from pathlib import Path

def get_added_images(
    acquisition: str, uuid: str, n_iter: int = 5
) -> Image:
    base_dir = Path("./models") / acquisition / uuid
    model_dirs = filter(lambda x: x.is_dir(), base_dir.iterdir())
```

```
 9      model_dirs = sorted(model_dirs, key=lambda x: int(x.stem))
10      imgs_per_label = get_imgs_per_label(model_dirs)
11      imgs = []
12      for i in range(10):
13          label_img = np.hstack(imgs_per_label[i])[:, -(28 * n_iter):]
14          imgs.append(label_img)
15      return Image.fromarray(np.vstack(imgs))
```

We can then call these functions, in our case with the following setup and *UUID*s:

```
1  uuid = "bc1adec5-bc34-44a6-a0eb-fa7cb67854e4"
2  img_random = get_added_images("random", uuid)
3  uuid = "5c8d6001-a5fb-45d3-a7cb-2a8a46b93d18"
4  img_ku = get_added_images("knowledge_uncertainty", uuid)
```

Let's compare the output.

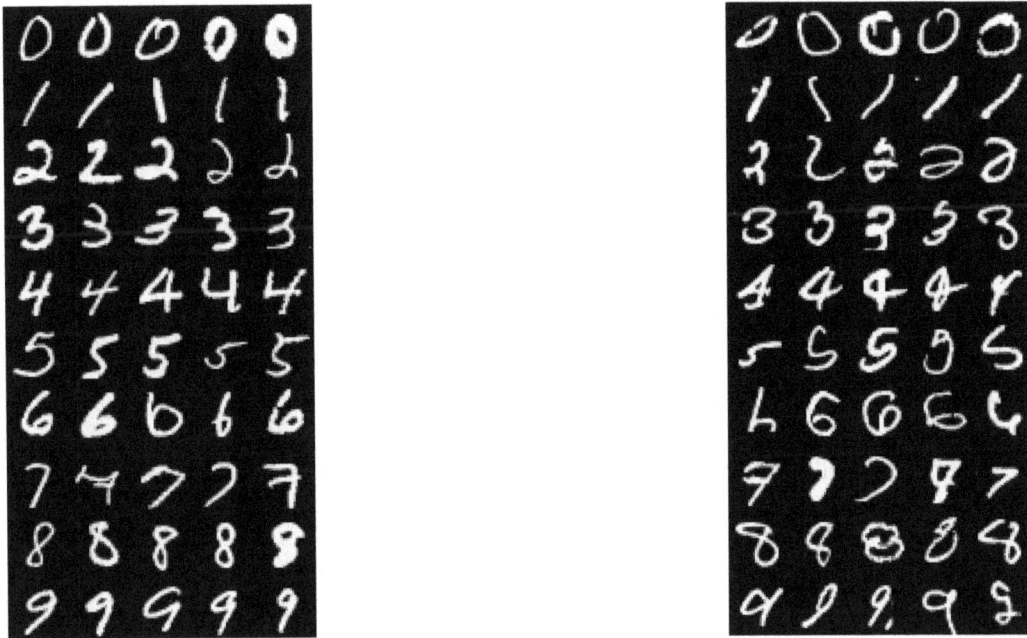

Figure 8.9: Images randomly selected (left) and images selected via knowledge uncertainty with MC dropout (right). Every row shows the last five images selected for the label

We can see in *Figure 8.9* that the images selected by the knowledge uncertainty acquisition function are probably more difficult to classify compared to the randomly selected images. The uncertainty acquisition function selects quite a few unusual representations of the digits in the dataset. Because our acquisition function was able to select these images, the model was better able to understand the full distribution of the dataset, which resulted in better accuracy over time.

8.5 Using uncertainty estimates for smarter reinforcement learning

Reinforcement learning aims to develop machine learning techniques capable of learning from their environment. There's a clue to the fundamental principle behind reinforcement learning in its name: the aim is to reinforce successful behaviour. Generally speaking, in reinforcement learning, we have an agent capable of executing a number of actions in an environment. Following these actions, the agent receives feedback from the environment, and this feedback is used to allow the agent to build a better understanding of which actions are more likely to lead to a positive outcome given the current state of the environment.

Formally, we can describe this using a set of states, S, a set of actions A, which map from a current state s to a new state s', and a reward function, $R(s, s')$, describing the reward for the transition between the current state, s, and the new state, s'. The set of states comprises a set of environment states, S_e, and a set of agent states, S_a, which together describe the state of the entire system.

We can think of this in terms of a game of Marco Polo, wherein call and response is used by one player in order to find another player. When the seeking player calls "Marco," the other player replies "Polo," giving the seeking player an estimate of their location based on the direction and amplitude of the sound. If we simplify this to consider it in terms of distance, a closer state would be one for which the distance reduces, such as $\delta = d - d' > 0$, where d is the distance at state s and d' is the distance for state s'. Conversely, a further state would be one for which $\delta = d - d' < 0$. Thus, for this example, we can use our δ values as feedback for our model, making our reward function $\delta = R(s, s') = d - d'$.

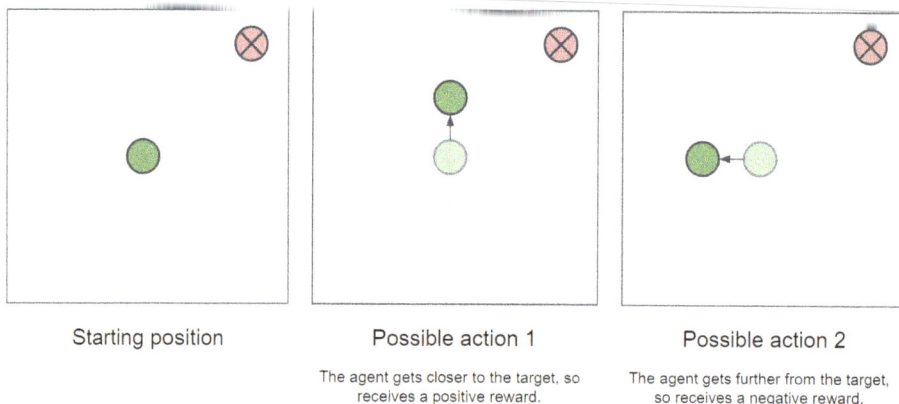

Starting position | Possible action 1 | Possible action 2

The agent gets closer to the target, so receives a positive reward.

The agent gets further from the target, so receives a negative reward.

Figure 8.10: Illustration of a Marco Polo reinforcement learning scenario

Let's consider our agent as the seeking player and its target as the hiding player. At each step, our agent collects more information about its environment, enabling it to better model the relationship between its actions $A(s)$ and the reward function $R(s, s')$ (in other words, it's learning which general direction it needs to move in to get closer to the target). At each step, we need to predict the reward function given the set of possible actions at the current state, A_s, so that we can choose the action that is most likely to maximize this reward function. In this case, the set of actions could be a set of directions we can move in, for example: forward, back, left, and right.

Traditional reinforcement learning uses a method called **Q Learning** to learn the relationship between the state, action, and reward. Q Learning doesn't involve neural network models, and instead accumulates state, action, and reward information in a table – the Q table – which is then used to determine the action most likely to produce the highest reward given the current state. While Q Learning is powerful, it becomes computationally prohibitive for large numbers of states and actions. To address this, researchers introduced the concept of **Deep Q Learning**, wherein the Q table is replaced by a neural network. Over a (usually large) number of iterations, the neural network learns which actions are likely to produce a higher reward given the current state.

To predict which action is likely to yield the highest reward value, we use a model trained on all historical actions, A_h, states S_h, and rewards, R_h. Our training input X comprises the actions A_h and states S_h, while our target output y comprises the reward values R_h. We can then use the model as part of a **Model Predictive Controller**, or **MPC**, which will select the action depending on which action is associated with the highest predicted reward:

$$a_{next} = \arg\max y_i \forall a_i \in A_s \tag{8.6}$$

Here, y_i is the reward prediction produced by our model, $f(a_i, s)$, which maps the current state s and possible actions $a_i \in A_s$ to reward values. However, before our model is of any use, we'll need to gather data to train on. We'll accrue data over a number of episodes, wherein each episode comprises a set of actions taken by the agent until some termination criteria are met. The ideal termination criterion would be the agent finding the target, but we can set other criteria, such as the agent encountering an obstacle, or the agent exhausting a maximum number of actions. Because the model has no information to start off with, we use a greedy policy commonly used in reinforcement learning, called an $\epsilon greedy$ policy, to allow the agent to start by randomly sampling from its environment. The idea here is that our agent will perform a random action with probability ϵ, and will otherwise use model predictions to select the action. After each episode, we will decrease ϵ, such that the agent will eventually be selecting actions based solely on the model. Let's put together a simple reinforcement learning example to see all of this in action.

Step 1: Initializing our environment

Our reinforcement learning example will be centred around our environment: this defines the space in which everything happens. We'll handle this with the `Environment` class. First, we set up our environment parameters:

```
1   import numpy as np
2   import tensorflow as tf
3   from scipy.spatial.distance import euclidean
4   from tensorflow.keras import (
5       Model,
6       Sequential,
7       layers,
8       optimizers,
9       metrics,
10      losses,
11  )
12  import pandas as pd
13  from sklearn.preprocessing import StandardScaler
14  import copy
15
```

```
16
17  class Environment:
18      def __init__(self, env_size=8, max_steps=2000):
19          self.env_size = env_size
20          self.max_steps = max_steps
21          self.agent_location = np.zeros(2)
22          self.target_location = np.random.randint(0, self.env_size, 2)
23          self.action_space = {
24              0: np.array([0, 1]),
25              1: np.array([0, -1]),
26              2: np.array([1, 0]),
27              3: np.array([-1, 0]),
28          }
29          self.delta = self.compute_distance()
30          self.is_done = False
31          self.total_steps = 0
32          self.ideal_steps = self.calculate_ideal_steps()
33      ...
```

Here, notice our environment size, denoted by env_size, which defines the number of rows and columns in our environment – in this case, we'll have an environment of size 8×8, resulting in 64 locations (for simplicity, we'll stick with a square environment). We'll also set a max_steps limit so that episodes don't go on too long while our agent is randomly selecting actions.

We also set the agent_location and target_location variables – the agent always starts at point [0, 0], while the target location is randomly allocated.

Next, we create a dictionary to map an integer value to an action. Going from 0 to 3, these actions are: forward, backward, right, left. We also set the delta variable – this is the initial distance between the agent and the target (we'll see how compute_distance() is implemented in a moment).

Finally, we initialize a few variables for tracking whether the termination criteria have been met (is_done), the total number of steps (total_steps), and the ideal number of steps (ideal_steps). The latter of these variables is the minimum number of steps required for the agent to get to the target from its starting position. We'll use this to calculate the regret, which is a useful indicator of performance for reinforcement learning and optimization algorithms. To calculate this, we'll add the following two functions to our class:

```
1       ...
2
3       def calculate_ideal_action(self, agent_location, target_location):
4           min_delta = 1e1000
5           ideal_action = -1
6           for k in self.action_space.keys():
7               delta = euclidean(
8                   agent_location + self.action_space[k], target_location
9               )
10              if delta <= min_delta:
11                  min_delta = delta
12                  ideal_action = k
13          return ideal_action, min_delta
14
15      def calculate_ideal_steps(self):
16          agent_location = copy.deepcopy(self.agent_location)
17          target_location = copy.deepcopy(self.target_location)
18          delta = 1e1000
19          i = 0
20          while delta > 0:
21              ideal_action, delta = self.calculate_ideal_action(
22                  agent_location, target_location
23              )
24              agent_location += self.action_space[ideal_action]
```

```
25              i += 1
26          return i
27      . . .
```

Here, `calculate_ideal_steps()` will run until the distance (`delta`) between the agent and the target is zero. At each iteration, it uses `calculate_ideal_action()` to select the action that will move the agent closest to the target.

Step 2: Updating the state of our environment

Now that we've initialized our environment, we need to add one of the most crucial pieces of our class: the `update` method. This controls what happens to our environment when the agent takes a new action:

```
1      . . .
2      def update(self, action_int):
3          self.agent_location = (
4              self.agent_location + self.action_space[action_int]
5          )
6          # prevent the agent from moving outside the bounds of the environment
7          self.agent_location[self.agent_location > (self.env_size - 1)] = (
8              self.env_size - 1
9          )
10         self.compute_reward()
11         self.total_steps += 1
12         self.is_done = (self.delta == 0) or (self.total_steps >= self.max_steps)
13         return self.reward
14     . . .
```

The method receives an action integer, and uses this to access the corresponding action in the `action_space` dictionary we defined earlier. It then updates the agent location. Because both the agent location and action are vectors, we can simply use vector addition to do this. Next, we check whether the agent has moved out of bounds of our environment

– if it has, we simply adjust its location so that it remains within our environment boundary.

The next line is another crucial piece of code: computing the reward with `compute_reward()` – we'll take a look at this in just a moment. Once we've computed the reward, we increment the `total_steps` counter, check our termination criteria, and return the reward value for the action.

We determine the reward using the following function. This will return a low reward (1) if the distance between the agent and the target increases, and a high reward (10) if the distance between the agent and target decreases:

```
1    . . .
2    def compute_reward(self):
3        d1 = self.delta
4        self.delta = self.compute_distance()
5        if self.delta < d1:
6            self.reward = 10
7        else:
8            self.reward = 1
9    . . .
```

This uses the `compute_distance()` function, which calculates the Euclidean distance between the agent and the target:

```
1    . . .
2    def compute_distance(self):
3        return euclidean(self.agent_location, self.target_location)
4    . . .
```

Lastly, we need a function to allow us to fetch the state of the environment, so that we can associate this with the reward values. We define this as follows:

```
1    . . .
2    def get_state(self):
3        return np.concatenate([self.agent_location, self.target_location])
4    . . .
```

Step 3: Defining our model

Now that we've set up our environment, we'll create a model class. This class will handle model training and inference, as well as selecting the best action according to the model's predictions. As always, we start with the __init__() method:

```
1    class RLModel:
2    def __init__(self, state_size, n_actions, num_epochs=500):
3        self.state_size = state_size
4        self.n_actions = n_actions
5        self.num_epochs = 200
6        self.model = Sequential()
7        self.model.add(
8            layers.Dense(
9                20, input_dim=self.state_size, activation="relu", name="layer_1"
10           )
11       )
12       self.model.add(layers.Dense(8, activation="relu", name="layer_2"))
13       self.model.add(layers.Dense(1, activation="relu", name="layer_3"))
14       self.model.compile(
15           optimizer=optimizers.Adam(),
16           loss=losses.Huber(),
17           metrics=[metrics.RootMeanSquaredError()],
18       )
19   . . .
```

Here, we pass a few variables related to our environment, such as the state size and number of actions. The code relating to the model definition should be familiar – we're simply instantiating a neural network using Keras. One point to note is that we're using the Huber loss here, instead of something more common such as the mean squared error. This is a common choice in robust regression tasks and in reinforcement learning. The reason for this is that the Huber loss dynamically switches between mean squared error and mean absolute error. The former is very good at penalizing small errors, while the latter is more robust to outliers. Through the Huber loss, we arrive at a loss function that is both robust to outliers and penalizes small errors.

This is particularly important in reinforcement learning because of the exploratory nature of the algorithms: we will often encounter some examples that are very exploratory, deviating significantly from the rest of the data, and thus causing large errors during training.

With our class initialization out of the way, we move on to our `fit()` and `predict()` functions:

```
1   ...
2       def fit(self, X_train, y_train, batch_size=16):
3           self.scaler = StandardScaler()
4           X_train = self.scaler.fit_transform(X_train)
5           self.model.fit(
6               X_train,
7               y_train,
8               epochs=self.num_epochs,
9               verbose=0,
10              batch_size=batch_size,
11          )
12
13
14      def predict(self, state):
15          rewards = []
16          X = np.zeros((self.n_actions, self.state_size))
17          for i in range(self.n_actions):
```

```
18                  X[i] = np.concatenate([state, [i]])
19              X = self.scaler.transform(X)
20              rewards = self.model.predict(X)
21              return np.argmax(rewards)
```

The `fit()` function should look very familiar – we're just scaling our inputs before fitting our Keras model. The `predict()` function has a little more going on. Because we need predictions for each of our possible actions (forward, backward, right, left), we need to generate inputs for these. We do so by concatenating the integer value associated with the action to the state, producing our complete state-action vector as we see on line 11. Doing this for all actions results in our input matrix, X, for which each row is associated with a specific action. We then scale X and run inference on this to obtain our predicted reward values. To select an action, we simply use `np.argmax()` to obtain the index associated with the highest predicted reward.

Step 4: Running our reinforcement learning

Now that we've defined our `Environment` and `RLModel` classes, we're ready to do some reinforcement learning! Let's first set up some important variables and instantiate our model:

```
1   env_size = 8
2   state_size = 5
3   n_actions = 4
4   epsilon = 1.0
5   history = {"state": [], "reward": []}
6   n_samples = 1000
7   max_steps = 500
8   regrets = []
9
10  model = RLModel(state_size, n_actions)
```

Most of these should be familiar by now, but we'll go over a few that we've not yet covered. The history dictionary is where we'll store our state and reward information as we progress through each step in each episode. We'll then use this information to train our model. Another unfamiliar variable here is n_samples – we're setting this because, rather than using all available data each time we train our model, we'll sample 1,000 data points from our data. This helps to avoid our training time exploding as we accrue more and more data. The last new variable here is regrets. This list will store our regret values for each episode. In our case, regret is defined simply as the difference between the number of steps taken by the model and the minimum number of steps required for the agent to reach the target:

$$regret = steps_{model} - steps_{ideal} \tag{8.7}$$

As such, regret is zero $\iff$ $steps_{model} == steps_{ideal}$. Regret is useful for measuring performance as our model learns, as we'll see in a moment. All that's left is the main loop of our reinforcement learning process:

```
for i in range(100):
    env = Environment(env_size, max_steps=max_steps)
    while not env.is_done:
        state = env.get_state()
        if np.random.rand() < epsilon:
            action = np.random.randint(n_actions)
        else:
            action = model.predict(state)
        reward = env.update(action)
        history["state"].append(np.concatenate([state, [action]]))
        history["reward"].append(reward)
    print(
        f"Completed episode {i} in {env.total_steps} steps."
        f"Ideal steps: {env.ideal_steps}."
```

```
15          f"Epsilon: {epsilon}"
16      )
17      regrets.append(np.abs(env.total_steps-env.ideal_steps))
18      idxs = np.random.choice(len(history["state"]), n_samples)
19      model.fit(
20          np.array(history["state"])[idxs],
21          np.array(history["reward"])[idxs]
22      )
23      epsilon-=epsilon/10
```

Here, we have our reinforcement learning process run from 100 episodes, reinitializing the environment each time. As we can see from the internal `while` loop, we will continue iterating – updating our agent and measuring our reward – until one of the termination criteria is met (either the agent reaches the target, or we run for the maximum allowed number of iterations).

After each episode, a `print` statement lets us know that the episode completed without error, and tells us how our agent did compared to the ideal number of steps. We then calculate the regret and append this to our `regrets` list, sample from our data in `history` and fit our model on the sampled data. Lastly, we finish each iteration of the outer loop by reducing epsilon.

After running this, we can additionally plot our regret values to see how we did:

```
1  import matplotlib.pyplot as plt
2  import seaborn as sns
3
4  df_plot = pd.DataFrame({"regret": regrets, "episode": np.arange(len(regrets))})
5  sns.lineplot(x="episode", y="regret", data=df_plot)
6  fig = plt.gcf()
7  fig.set_size_inches(5, 10)
8  plt.show()
```

This produces the following plot, showing how our model did over the 100 episodes:

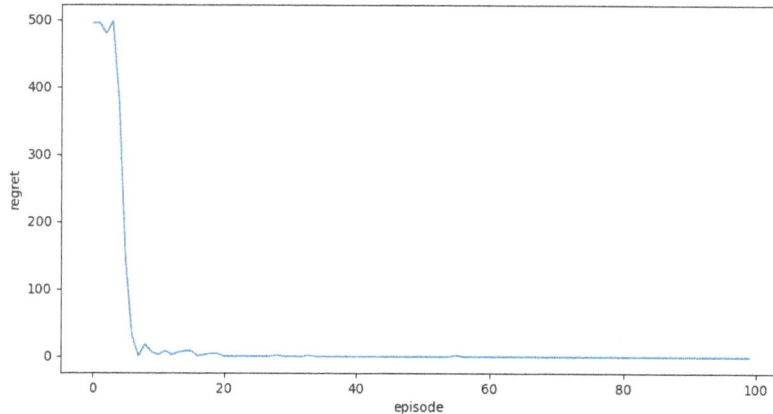

Figure 8.11: Plot of regret values following 100 episodes of reinforcement learning

As we can see here, it did poorly to begin with, but the model quickly learned to predict reward values, allowing it to predict optimal actions, and reducing regret to 0.

So far, things are pretty simple. In fact, you may be wondering why we need a model at all – why not just calculate the distance between the target and the proposed positions, and select an action accordingly? Well, firstly, the aim of reinforcement learning is for an agent to discover how to interact in a given setting without any prior knowledge so while our agent can execute actions, it has no concept of distance. This is something that is learned through interacting with the environment. Secondly, it may not be that simple: what if there are obstacles in the environment? In this case, our agent needs to be more intelligent than simply moving toward the sound.

While this is just an illustrative example, real-world applications of reinforcement learning involve scenarios for which we have very limited knowledge, and thus designing an agent that can explore its environment and learn how to interact optimally allows us to develop models for applications for which supervised methods aren't an option.

Another factor to consider in real-world scenarios is risk: we want our agent to make *sensible* decisions, not just decisions that maximize the reward: we need it to build some understanding of the risk/reward trade-off. This is where uncertainty estimates come in.

8.5.1 Navigating obstacles with uncertainty

With uncertainty estimates, we can balance the reward against the model's confidence in its prediction. If its confidence is low (meaning that uncertainty is high), then we may want to be cautious about how we incorporate our model's predictions. For example, let's take the reinforcement learning scenario we've just explored. For each episode, our model is predicting which action will yield the highest reward, and our agent then chooses this action. In the real world, things aren't so predictable – our environment can change, leading to unexpected consequences. What if an obstacle appears in our environment, and colliding with that obstacle prevents our agent from completing its task? Well, clearly if our agent hasn't yet encountered the obstacle, it's doomed to fail. Fortunately, in the case of Bayesian Deep Learning, this isn't the case. As long as we have some way of sensing the obstacle, our agent can detect the obstacle and take a different route – even if the obstacle wasn't encountered in previous episodes.

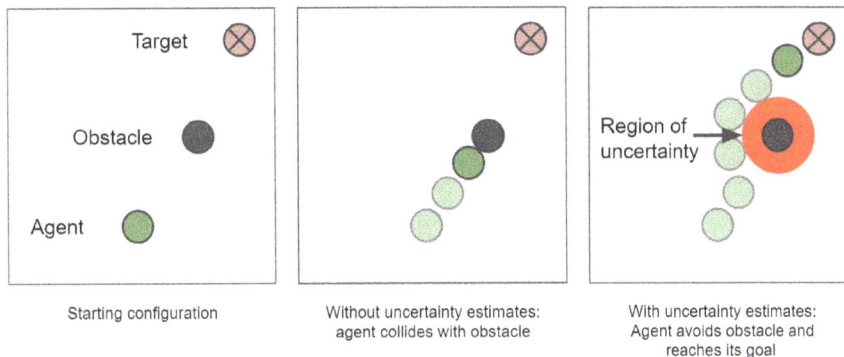

Figure 8.12: Illustration of how uncertainty affects the actions of a reinforcement learning agent

This is possible thanks to our uncertainty estimates. When the model encounters something unusual, its uncertainty estimate for that prediction will be high. Thus, if we incorporate this into our MPC equation, we can balance reward with uncertainty, ensuring that we prioritize lower risk over higher reward. To do so, we modify our MPC equation as follows:

$$a_{next} = \arg\max(y_i - \lambda\sigma_i)\forall a_i \in A_s \qquad (8.8)$$

Here, we see that we're now subtracting a value, $\lambda\sigma_i$, from our reward prediction y_i. This is because σ_i is our uncertainty associated with the ith prediction. We use λ to scale the uncertainty so that it appropriately penalizes uncertain actions; this is a parameter we can tune depending on the application. With a sufficiently well calibrated method, we'll see larger values for σ_i in cases where the model is uncertain about its predictions. Let's build on our earlier code example to see this in action.

Step 1: Introducing obstacles

To create a challenge for our agent, we're going to introduce obstacles to our environment. To test how our agent responds to unfamiliar input, we're going to change the policy that our obstacle follows - it will either follow a static policy or a dynamic policy depending on our environment settings. We'll change the __init__() function for our Environment class to incorporate these changes:

```
1    def __init__(self, env_size=8, max_steps=2000, dynamic_obstacle=False, lambda_va
2        self.env_size = env_size
3        self.max_steps = max_steps
4        self.agent_location = np.zeros(2)
5        self.dynamic_obstacle = dynamic_obstacle
6        self.lambda_val = lambda_val
7        self.target_location = np.random.randint(0, self.env_size, 2)
8        while euclidean(self.agent_location, self.target_location) < 4:
9            self.target_location = np.random.randint(0, self.env_size, 2)
10       self.action_space = {
11           0: np.array([0, 1]),
12           1: np.array([0, -1]),
13           2: np.array([1, 0]),
```

```
14                3: np.array([-1, 0]),
15            }
16            self.delta = self.compute_distance()
17            self.is_done = False
18            self.total_steps = 0
19            self.obstacle_location = np.array(
20                [self.env_size / 2, self.env_size / 2], dtype=int
21            )
22            self.ideal_steps = self.calculate_ideal_steps()
23            self.collision = False
24
```

There's quite a lot going on here, so we'll go through each of the changes. First, to determine whether the obstacle is static or dynamic, we set the dynamic_obstacle variable. If this is **True**, then we'll randomly set the obstacle location. If it's **False**, then our object will sit in the middle of our environment. We're also setting our **lambda** (λ) parameter here, which defaults to 2.

We've also introduced a **while** loop here when setting target_location: we've done this to ensure that there's some distance between the agent and the target. We need to do this to ensure there's space between our agent and our target to drop in our dynamic obstacle – otherwise our agent may never encounter the obstacle (which would somewhat defeat the point of this example).

Lastly, we compute the obstacle location on line 17: you'll note that this just sets it to the middle of the environment. This is because we use the dynamic_obstacle flag later on to place the obstacle between the agent and the target – we do this during the calculate_ideal_steps() function, as this way we know the obstacle will lie along the agent's ideal path (and is thus more likely to be encountered).

Step 2: Placing our dynamic obstacle

When dynamic_obstacle is **True**, we want to place our obstacle somewhere different each episode, thus posing more of a challenge for our agent. To do so, we add a modification to the calculate_ideal_steps() function, as mentioned previously:

```
def calculate_ideal_steps(self):
    agent_location = copy.deepcopy(self.agent_location)
    target_location = copy.deepcopy(self.target_location)
    delta = 1e1000
    i = 0
    while delta > 0:
        ideal_action, delta = self.calculate_ideal_action(
            agent_location, target_location
        )
        agent_location += self.action_space[ideal_action]
        if np.random.randint(0, 2) and self.dynamic_obstacle:
            self.obstacle_location = copy.deepcopy(agent_location)
        i += 1
    return i
```

Here, we see that we call np.random.randint(0, 2) on each iteration of the **while** loop. This is to randomize in which position the obstacle is placed along the ideal path.

Step 3: Adding sensing

Our agent will have no hope of avoiding the object introduced into our environment if it can't sense the object. As such, we'll add a function to simulate a sensor: get_obstacle_proximity This sensor will give our agent information on how close it would get to an object were it to take a certain action. We'll have this return progressively higher values depending on how close to the object a given action would place our agent. If the action places our agent sufficiently far from the object (in this case, at least 4.5 spaces), then our sensor will return zero. This sensing function allows our agent to effectively see one step ahead, so we can think of the sensor as having a range of one step.

```
1    def get_obstacle_proximity(self):
2        obstacle_action_dists = np.array(
3            [
4                euclidean(
5                    self.agent_location + self.action_space[k],
6                    self.obstacle_location,
7                )
8                for k in self.action_space.keys()
9            ]
10       )
11       return self.lambda_val * (
12           np.array(obstacle_action_dists < 2.5, dtype=float)
13           + np.array(obstacle_action_dists < 3.5, dtype=float)
14           + np.array(obstacle_action_dists < 4.5, dtype=float)
15       )
```

Here, we first compute the future proximity for the agent given each action, after which we compute integer "proximity" values. These are computed by first constructing Boolean arrays for each proximity condition, in this case being $\delta_o < 2.5$, $\delta_o < 3.5$, and $\delta_o < 4.5$, where δ_o is the distance to the obstacle. We then sum these such that the proximity score has integer values of 3, 2, or 1 depending on how many of the criteria are met. This gives us a sensor that returns some basic information about the obstacle's future proximity for each of the proposed actions.

Step 4: Modifying our reward function

The last thing we need to do to prepare our environment is to update our reward function:

```
1    def compute_reward(self):
2        d1 = self.delta
3        self.delta = self.compute_distance()
4        if euclidean(self.agent_location, self.obstacle_location) == 0:
```

```
5              self.reward = 0
6              self.collision = True
7              self.is_done = True
8          elif self.delta < d1:
9              self.reward = 10
10         else:
11             self.reward = 1
```

Here, we've added a statement to check whether the agent and obstacle have collided (checking whether the distance between the two is zero). If so, we'll return a reward of 0, and set both the `collision` and `is_done` variables to `True`. This introduces a new termination criteria, **collision**, and will allow our agent to learn that collisions are bad, as these receive the lowest reward.

Step 5: Initializing our uncertainty-aware model

Now that our environment is ready, we need a new model – one capable of producing uncertainty estimates. For this model, we'll use an MC dropout network with a single hidden layer:

```
1  class RLModelDropout:
2      def __init__(self, state_size, n_actions, num_epochs=200, nb_inference=10):
3          self.state_size = state_size
4          self.n_actions = n_actions
5          self.num_epochs = num_epochs
6          self.nb_inference = nb_inference
7          self.model = Sequential()
8          self.model.add(
9              layers.Dense(
10                 10, input_dim=self.state_size, activation="relu", name="layer_1"
11             )
12         )
13         # self.model.add(layers.Dropout(0.15))
```

```
14          # self.model.add(layers.Dense(8, activation='relu', name='layer_2'))
15          self.model.add(layers.Dropout(0.15))
16          self.model.add(layers.Dense(1, activation="relu", name="layer_2"))
17          self.model.compile(
18              optimizer=optimizers.Adam(),
19              loss=losses.Huber(),
20              metrics=[metrics.RootMeanSquaredError()],
21          )
22
23          self.proximity_dict = {"proximity sensor value": [], "uncertainty": []}
24      ...
```

This should look pretty familiar, but you'll notice a few key differences. First, we're again using the Huber loss. Secondly, we've introduced a dictionary, proximity_dict, which will record the proximity values received from the sensor and the associated model uncertainties. This will allow us to evaluate our model's sensitivity to anomalous proximity values later on.

Step 6: Fitting our MC dropout network

Next, we need the following lines:

```
1       ...
2       def fit(self, X_train, y_train, batch_size=16):
3           self.scaler = StandardScaler()
4           X_train = self.scaler.fit_transform(X_train)
5           self.model.fit(
6               X_train,
7               y_train,
8               epochs=self.num_epochs,
9               verbose=0,
10              batch_size=batch_size,
11          )
```

```
12      . . .
```

This should again look very familiar – we're simply preparing our data by first scaling our inputs before fitting our model.

Step 7: Making predictions

Here, we see that we've slightly modified our `predict()` function:

```
1      . . .
2      def predict(self, state, obstacle_proximity, dynamic_obstacle=False):
3          rewards = []
4          X = np.zeros((self.n_actions, self.state_size))
5          for i in range(self.n_actions):
6              X[i] = np.concatenate([state, [i], [obstacle_proximity[i]]])
7          X = self.scaler.transform(X)
8          rewards, y_std = self.predict_ll_dropout(X)
9          # we subtract our standard deviations from our predicted reward values,
10         # this way uncertain predictions are penalised
11         rewards = rewards - (y_std * 2)
12         best_action = np.argmax(rewards)
13         if dynamic_obstacle:
14             self.proximity_dict["proximity sensor value"].append(
15                 obstacle_proximity[best_action]
16             )
17             self.proximity_dict["uncertainty"].append(y_std[best_action][0])
18         return best_action
19     . . .
```

More specifically, we've added the `obstacle_proximity` and `dynamic_obstacle` variables. The former allows us to receive the sensor information and incorporate this in the inputs we pass to our model. The latter is a flag telling us whether we're in the dynamic obstacle phase – if so, we want to record information about the sensor values and uncertainties in our `proximity_dict` dictionary.

The next block of prediction code should again look very familiar:

```
1   ...
2       def predict_ll_dropout(self, X):
3           ll_pred = [
4               self.model(X, training=True) for _ in range(self.nb_inference)
5           ]
6           ll_pred = np.stack(ll_pred)
7           return ll_pred.mean(axis=0), ll_pred.std(axis=0)
```

This function simply implements the MC dropout inference, obtaining predictions for
nb_inference forward passes, and returns the means and standard deviations associated
with our predictive distributions.

Step 8: Adapting our standard model

To understand the difference that our Bayesian model makes, we'll need to compare it
with a non-Bayesian model. As such, we'll update our RLModel class from earlier, adding
the ability to incorporate proximity information from our proximity sensor:

```
1   class RLModel:
2       def __init__(self, state_size, n_actions, num_epochs=500):
3           self.state_size = state_size
4           self.n_actions = n_actions
5           self.num_epochs = 200
6           self.model = Sequential()
7           self.model.add(
8               layers.Dense(
9                   20, input_dim=self.state_size, activation="relu", name="layer_1"
10              )
11          )
12          self.model.add(layers.Dense(8, activation="relu", name="layer_2"))
13          self.model.add(layers.Dense(1, activation="relu", name="layer_3"))
```

```
14          self.model.compile(
15              optimizer=optimizers.Adam(),
16              loss=losses.Huber(),
17              metrics=[metrics.RootMeanSquaredError()],
18          )
19
20      def fit(self, X_train, y_train, batch_size=16):
21          self.scaler = StandardScaler()
22          X_train = self.scaler.fit_transform(X_train)
23          self.model.fit(
24              X_train,
25              y_train,
26              epochs=self.num_epochs,
27              verbose=0,
28              batch_size=batch_size,
29          )
30
31      def predict(self, state, obstacle_proximity, obstacle=False):
32          rewards = []
33          X = np.zeros((self.n_actions, self.state_size))
34          for i in range(self.n_actions):
35              X[i] = np.concatenate([state, [i], [obstacle_proximity[i]]])
36          X = self.scaler.transform(X)
37          rewards = self.model.predict(X)
38          return np.argmax(rewards)
39
```

Crucially, we see here that our decision function has not changed: because we don't have model uncertainties, our model's predict() function is choosing actions based only on the predicted reward.

Step 9: Preparing to run our new reinforcement learning experiment

Now we're ready to set up our new experiment. We'll initialize the variables we used previously, and will introduce a few more:

```
1   env_size = 8
2   state_size = 6
3   n_actions = 4
4   epsilon = 1.0
5   history = {"state": [], "reward": []}
6   model = RLModelDropout(state_size, n_actions, num_epochs=400)
7   n_samples = 1000
8   max_steps = 500
9   regrets = []
10  collisions = 0
11  failed = 0
```

Here, we see that we've introduced a `collisions` variable and a `failed` variable. These will keep track of the number of collisions and the number of failed episodes so that we can compare the performance of our Bayesian model with that of our non-Bayesian model. We're now ready to run our experiment!

Step 10: Running our BDL reinforcement experiment

As before, we're going to run our experiment for 100 episodes. However, this time, we're only going to run training on our model for the first 50 episodes. After that, we'll stop training, and evaluate how well our model is able to find a safe path to the target. During these last 50 episodes, we'll set `dynamic_obstacle` to `True`, meaning our environment will now randomly choose a different position for our obstacle for each episode. Importantly, these random positions will be *along the ideal path* between the agent and its target.

Let's take a look at the code:

```
1   for i in range(100):
2       if i < 50:
3           env = Environment(env_size, max_steps=max_steps)
4           dynamic_obstacle = False
5       else:
6           dynamic_obstacle = True
7           epsilon = 0
8           env = Environment(
9               env_size, max_steps=max_steps, dynamic_obstacle=True
10          )
11      ...
```

First, we check whether the episode is within the first 50 episodes. If so, we instantiate our environment with dynamic_obstacle=False, and also set our global dynamic_obstacle variable to False.

If the episode is one of the last 50 episodes, we create an environment with a randomly placed obstacle, and also set epsilon to 0, to ensure we're always using our model predictions when selecting actions.

Next, we enter our while loop, setting our agent in motion. This is very similar to the loop we saw in the last example, except this time we're calling env.get_obstacle_proximity(), using the returned obstacle proximity information in our predictions, and also storing this information in our episode history:

```
1       ...
2       while not env.is_done:
3           state = env.get_state()
4           obstacle_proximity = env.get_obstacle_proximity()
5           if np.random.rand() < epsilon:
6               action = np.random.randint(n_actions)
```

```
7       else:
8           action = model.predict(state, obstacle_proximity, dynamic_obstacle)
9       reward = env.update(action)
10      history["state"].append(
11          np.concatenate([state, [action],
12          [obstacle_proximity[action]]])
13      )
14      history["reward"].append(reward)
15      ...
```

Lastly, we'll record some information about completed episodes and print the outcome of the most recent episode to our terminal. We update our `failed` and `collisions` variables and print whether the episode was complete successfully, the agent failed to find the target, or the agent collided with the obstacle:

```
1    if env.total_steps == max_steps:
2        print(f"Failed to find target for episode {i}. Epsilon: {epsilon}")
3        failed += 1
4    elif env.total_steps < env.ideal_steps:
5        print(f"Collided with obstacle during episode {i}. Epsilon: {epsilon}")
6        collisions += 1
7    else:
8        print(
9            f"Completed episode {i} in {env.total_steps} steps."
10           f"Ideal steps: {env.ideal_steps}."
11           f"Epsilon: {epsilon}"
12       )
13   regrets.append(np.abs(env.total_steps-env.ideal_steps))
14   if not dynamic_obstacle:
15       idxs = np.random.choice(len(history["state"]), n_samples)
16       model.fit(
17           np.array(history["state"])[idxs],
18           np.array(history["reward"])[idxs]
```

```
19            )
20            epsilon-=epsilon/10
```

The last statement here also checks whether we're in the dynamic obstacle phase, and if not, runs a round of training and decrements our epsilon value (as with the last example).

So, how did we do? Repeating the above 100 episodes for both the RLModel and RLModelDropout models, we obtain the following results:

Model	Failed episodes	Collisions	Successful episodes
RLModelDropout	19	3	31
RLModel	16	10	34

Figure 8.13: A table showing collision predictions

As we see here, there are advantages and disadvantages to consider when choosing whether to use a standard neural network or a Bayesian neural network – the standard neural network achieves a greater number of successfully completed episodes. However, crucially, the agent using the Bayesian neural network only collided with the obstacle three times, compared to the standard method's 10 times – that's a 70% reduction in collisions!

Note that as the experiment is stochastic, your results may differ, but on the GitHub repository we have included the experiment complete with the seed used to produce these results.

We can get a better idea of why this is by looking at the data we recorded in RLModelDropout's proximity_dict dictionary:

```
1  import matplotlib.pyplot as plt
2  import seaborn as sns
3
```

```
4   df_plot = pd.DataFrame(model.proximity_dict)
5   sns.boxplot(x="proximity sensor value", y="uncertainty", data=df_plot)
```

This produces the following plot:

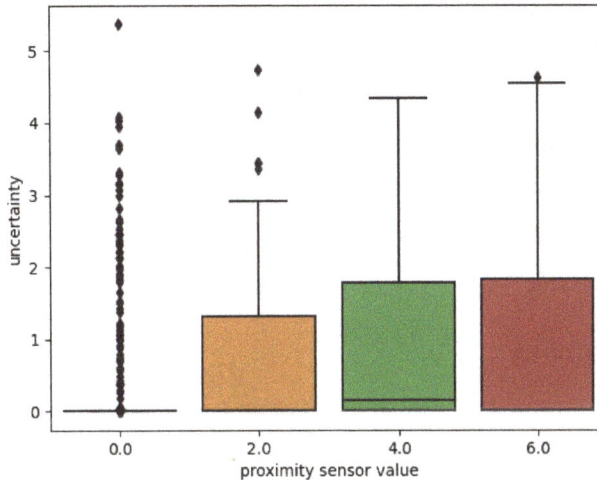

Figure 8.14: Distribution of uncertainty estimates associated with increasing proximity sensor values

As we see here, the model uncertainty estimates increase as the sensor values increase. This is because, during the first 50 episodes, our agent learns to avoid the centre of the environment (as this is where the obstacle is) – thus it gets used to low (or zero) proximity sensor values. This means that higher sensor values are anomalous, and are thus able to be picked up by the model's uncertainty estimates. Our agent then successfully accounts for this uncertainty using the uncertainty-aware MPC equation.

In this example, we saw how BDL can be applied to reinforcement learning to facilitate more cautious behaviour on the part of our reinforcement learning agents. While the example here was fairly basic, the implications are pretty significant: imagine this being applied to safety-critical applications. In these settings, we're often happy to accept poorer overall model performance if it meets better safety requirements. Thus, BDL has an important place within the field of safe reinforcement learning, allowing the development of reinforcement learning methods suitable for safety-critical scenarios.

In the next section, we'll see how BDL can be used to create models that are robust to another key consideration for real-world applications: adversarial inputs.

8.6 Susceptibility to adversarial input

In *Chapter 3*, we saw that we could fool a CNN by slightly perturbing the input pixels of an image. A picture that clearly looked like a cat was predicted as a dog with high confidence. The adversarial attack that we created (*FSGM*) is one of the many adversarial attacks that exist, and BDL might offer some protection against these attacks. Let's see how that works in practice.

Step 1: Model training

Instead of using a pre-trained model, as in *Chapter 3*, we train a model from scratch. We use the same train and test data from *Chapter 3* – see that chapter for instructions on how to load the dataset. As a reminder, the dataset is a relatively small dataset of cats and dogs. We first define our model. We use a VGG-like architecture but add dropout after every MaxPooling2D layer:

```
1  def conv_block(filters):
2      return [
3          tf.keras.layers.Conv2D(
4              filters,
5              (3, 3),
6              activation="relu",
7              kernel_initializer="he_uniform",
8          ),
9          tf.keras.layers.MaxPooling2D((2, 2)),
10         tf.keras.layers.Dropout(0.5),
11     ]
```

```
12
13
14  model = tf.keras.models.Sequential(
15      [
16          tf.keras.layers.Conv2D(
17              32,
18              (3, 3),
19              activation="relu",
20              input_shape=(160, 160, 3),
21              kernel_initializer="he_uniform",
22          ),
23          tf.keras.layers.MaxPooling2D((2, 2)),
24          tf.keras.layers.Dropout(0.2),
25          *conv_block(64),
26          *conv_block(128),
27          *conv_block(256),
28          *conv_block(128),
29          tf.keras.layers.Conv2D(
30              64,
31              (3, 3),
32              activation="relu",
33              kernel_initializer="he_uniform",
34          ),
35          tf.keras.layers.Flatten(),
36          tf.keras.layers.Dense(64, activation="relu"),
37          tf.keras.layers.Dropout(0.5),
38          tf.keras.layers.Dense(2),
39      ]
40  )
41
42
```

We then normalize our data, and compile and train our model:

```
1  train_dataset_preprocessed = train_dataset.map(lambda x, y: (x / 255., y))
2  val_dataset_preprocessed = validation_dataset.map(lambda x, y: (x / 255., y))
3
4  model.compile(optimizer=tf.keras.optimizers.Adam(learning_rate=0.001),
5                  loss=tf.keras.losses.CategoricalCrossentropy(from_logits=True),
6                  metrics=['accuracy'])
7  model.fit(
8      train_dataset_preprocessed,
9      epochs=200,
10     validation_data=val_dataset_preprocessed,
11 )
```

This will give us a model accuracy of about 85%.

Step 2: Running inference and evaluating our standard model

Now that we have trained our model, let's see how much protection it offers against an adversarial attack. In *Chapter 3*, we created an adversarial attack from scratch. In this chapter, we'll use the cleverhans library to create the same attack in one line for multiple images at once:

```
1  from cleverhans.tf2.attacks.fast_gradient_method import (
2      fast_gradient_method as fgsm,
3  )
```

Let's first measure the accuracy of our deterministic model on the original images and the adversarial images:

```
1  predictions_standard, predictions_fgsm, labels = [], [], []
2  for imgs, labels_batch in test_dataset:
3    imgs /= 255.
```

```
4    predictions_standard.extend(model.predict(imgs))
5    imgs_adv = fgsm(model, imgs, 0.01, np.inf)
6    predictions_fgsm.extend(model.predict(imgs_adv))
7    labels.extend(labels_batch)
```

Now that we have our predictions, we can print the accuracy:

```
1    accuracy_standard = CategoricalAccuracy()(
2        labels, predictions_standard
3    ).numpy()
4    accuracy_fgsm = CategoricalAccuracy()(
5        labels, predictions_fgsm
6    ).numpy()
7    print(f"{accuracy_standard=.2%}, {accuracy_fsgm=:.2%}")
8    # accuracy_standard=83.67%, accuracy_fsgm=30.70%
```

We can see that our standard model offers little protection against this adversarial attack. Although it performs pretty well on standard images, it has an accuracy of 30.70% on adversarial images! Let's see if a Bayesian model can do better. Because we trained our model with dropout, we can easily make it an MC dropout model. We create an inference function where we keep dropout at inference, as indicated by the training=**True** parameter:

```
1    import numpy as np
2
3
4    def mc_dropout(model, images, n_inference: int = 50):
5      return np.swapaxes(np.stack([
6          model(images, training=True) for _ in range(n_inference)
7      ]), 0, 1)
```

With this function in place, we can replace the standard loop with MC dropout inference. We keep track of all our predictions again and run inference on our standard images and the adversarial images:

```
1  predictions_standard_mc, predictions_fgsm_mc, labels = [], [], []
2  for imgs, labels_batch in test_dataset:
3      imgs /= 255.
4      predictions_standard_mc.extend(
5          mc_dropout(model, imgs, 50)
6      )
7      imgs_adv = fgsm(model, imgs, 0.01, np.inf)
8      predictions_fgsm_mc.extend(
9          mc_dropout(model, imgs_adv, 50)
10     )
11     labels.extend(labels_batch)
```

And we can again print our accuracy:

```
1  accuracy_standard_mc = CategoricalAccuracy()(
2      labels, np.stack(predictions_standard_mc).mean(axis=1)
3  ).numpy()
4  accuracy_fgsm_mc = CategoricalAccuracy()(
5      labels, np.stack(predictions_fgsm_mc).mean(axis=1)
6  ).numpy()
7  print(f"{accuracy_standard_mc=.2%}, {accuracy_fgsm_mc=:.2%}")
8  # accuracy_standard_mc=86.60%, accuracy_fgsm_mc=80.75%
```

We can see that our simple modification made the model setup much more robust to adversarial examples. Instead of an accuracy of around 30%, we now obtain accuracy of more than 80%, pretty close to the accuracy of 83% of the deterministic model on the non-perturbed images. Moreover, we can see that MC dropout also improves the accuracy on our standard images by a few percentage points, from 83% to 86%. Almost no method offers perfect robustness to adversarial examples, so the fact that we can get so close to our model's standard accuracy is a great achievement.

Because our model has not seen the adversarial images before, a model with good uncertainty values should also have a lower confidence on average on the adversarial images compared to a standard model. Let's see if this is the case. We create a function to compute the average softmax value of the predictions of our deterministic model and create a similar function for our MC dropout predictions:

```python
def get_mean_softmax_value(predictions) -> float:
    mean_softmax = tf.nn.softmax(predictions, axis=1)
    max_softmax = np.max(mean_softmax, axis=1)
    mean_max_softmax = max_softmax.mean()
    return mean_max_softmax


def get_mean_softmax_value_mc(predictions) -> float:
    predictions_np = np.stack(predictions)
    predictions_np_mean = predictions_np.mean(axis=1)
    return get_mean_softmax_value(predictions_np_mean)
```

We can then print the mean softmax score for both models:

```python
mean_standard = get_mean_softmax_value(predictions_standard)
mean_fgsm = get_mean_softmax_value(predictions_fgsm)
mean_standard_mc = get_mean_softmax_value_mc(predictions_standard_mc)
mean_fgsm_mc = get_mean_softmax_value_mc(predictions_fgsm_mc)
print(f"{mean_standard=:.2%}, {mean_fgsm=:.2%}")
print(f"{mean_standard_mc=:.2%}, {mean_fgsm_mc=:.2%}")
# mean_standard=89.58%, mean_fgsm=89.91%
# mean_standard_mc=89.48%, mean_fgsm_mc=85.25%
```

We can see that our standard model is actually slightly more confident on adversarial images compared to standard images, although its accuracy dropped significantly. However, our MC dropout model shows a lower confidence on the adversarial images compared to the standard images. Although the drop in confidence is not very large, it is good to see that the model is dropping its mean confidence on adversarial images, while keeping a reasonable accuracy.

8.7 Summary

In this chapter, we have illustrated the various applications of modern BDL in five different case studies. Each case study used code examples to highlight a particular strength of BDL in response to various, common problems in applied machine learning practice. First, we saw how BDL can be used to detect out-of-distribution images in a classification task. We then looked at how BDL methods can be used to make models more robust to dataset shift, which is a very common problem in production environments. Next, we learned how BDL can help us to select the most informative data points for training and updating our machine learning models. We then turned to reinforcement learning and saw how BDL can be used to facilitate more cautious behaviour in reinforcement learning agents. Finally, we saw how BDL can help us in the face of adversarial attacks.

In the next chapter, we will have a look at the future of BDL by reviewing current trends and the latest methods.

8.8 Further reading

The following reading list will offer a greater understanding of some of the topics we touched on in this chapter:

- *Benchmarking neural network robustness to common corruptions and perturbations*, Dan Hendrycks and Thomas Dietterich, 2019: this is the paper that introduced the image quality perturbations to benchmark model robustness, which we saw in the robustness case study.

- *Can You Trust Your Model's Uncertainty? Evaluating Predictive Uncertainty Under*

Dataset Shift, Yaniv Ovadia, Emily Fertig *et al.*, 2019: this comparison paper uses image quality perturbations to introduce artificial dataset shift at different severity levels and measures how different deep neural networks respond to dataset shift in terms of accuracy and calibration.

- *A Baseline for Detecting Misclassified and Out-of-Distribution Examples in Neural Networks*, Dan Hendrycks and Kevin Gimpel, 2016: this fundamental out-of-distribution detection paper introduces the concept and shows that softmax values are not perfect when it comes to OOD detection.

- *Enhancing The Reliability of Out-of-distribution Image Detection in Neural Networks*, Shiyu Liang, Yixuan Li and R. Srikant, 2017: shows that input perturbation and temperature scaling can improve the softmax baseline for out-of-distribution detection.

- *A Simple Unified Framework for Detecting Out-of-Distribution Samples and Adversarial Attacks*, Kimin Lee, Kibok Lee, Honglak Lee and Jinwoo Shin, 2018: shows that using the Mahalanobis distance can be effective for out-of-distribution detection.

9

Next Steps in Bayesian Deep Learning

Throughout this book, we've covered the fundamental concepts behind Bayesian deep learning (BDL), from understanding what uncertainty is and its role in developing robust machine learning systems, right through to learning how to implement and analyze the performance of several fundamental BDL. While what you've learned will equip you to start developing your own BDL solutions, the field is moving quickly, and there are many new techniques on the horizon.

To wrap up the book, in this chapter we'll take a look at the current trends in BDL, before we dive into some of the latest developments in the field. We'll conclude by introducing some alternatives to BDL, and provide some advice on additional resources you can use to continue your journey into Bayesian machine learning methods.

We'll cover the following sections:

- Current trends in BDL

- How are BDL methods being applied to solve real-world problems?

- Latest methods in BDL

- Alternatives to BDL

- Your next steps in BDL

9.1   Current trends in BDL

In this section, we'll explore the current trends in BDL. We'll look at which models are particularly popular in the literature and discuss why certain models have been selected for certain applications. This should give you a good impression of how the fundamentals covered throughout the book apply more broadly across a variety of application domains.

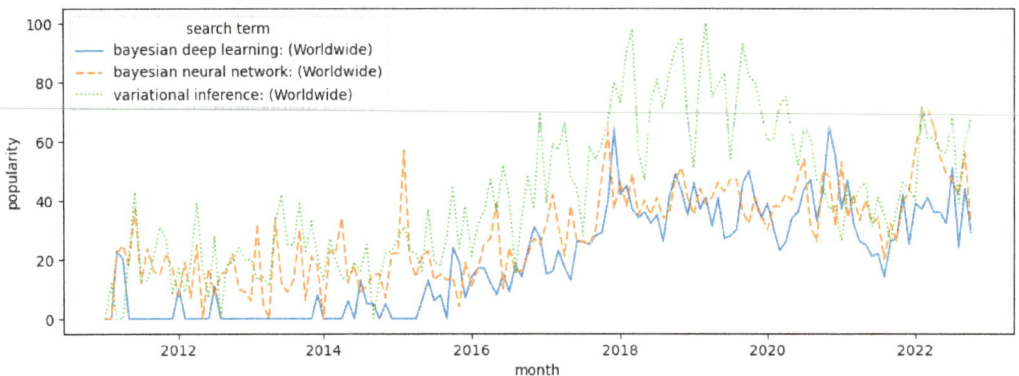

Figure 9.1: Popularity of key BDL search terms over time

As we see in *Figure* 9.1, there's been a marked increase in popularity of search terms related to BDL over the past decade. Unsurprisingly, this follows the trend in the popularity of deep learning search terms, as we see in *Figure 9.2*; as deep learning has become more popular, there has been increased interest in quantifying the uncertainty associated with the predictions produced by DNNs. Interestingly, these plots both show a similar dip in popularity in mid-late 2021, indicating that as long as deep learning is popular, there will also be interest in BDL.

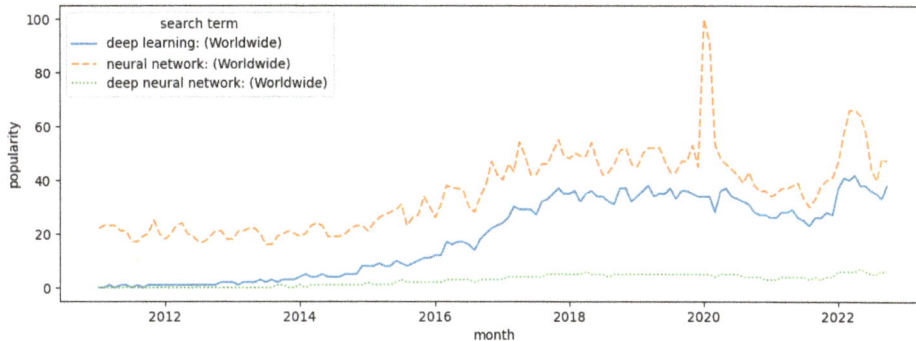

Figure 9.2: Popularity of key deep learning search terms over time

Figure 9.1 demonstrates another interesting point in that, generally speaking, the term *variational inference* is more popular than the other two BDL-related search terms we've used here. As mentioned in *Chapter 5* when we covered variational autoencoders, variational inference is one component of BDL that has made significant waves in the machine learning community, now being a feature of many different deep learning architectures. As such, it's no surprise that it is generally more popular than the terms that explicitly include the word "Bayesian."

But where do the methods that we've explored in the book fit in terms of their popularity and integration into a wide variety of deep learning solutions? We can learn more about this simply by looking at the citations for each of the original papers.

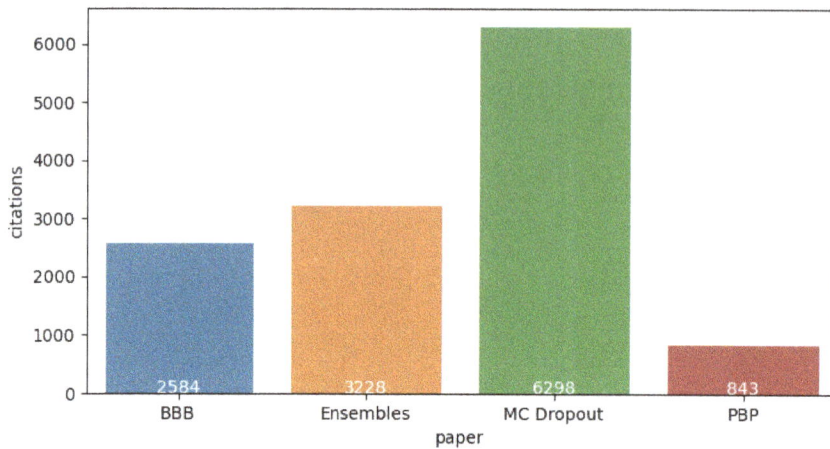

Figure 9.3: Popularity of key deep learning search terms over time

In *Figure 9.3*, we see that the MC dropout paper is by far the most popular paper when it comes to citations – coming in at nearly double that of the next most popular method. At this point in the book, the reasons for this should be fairly clear: not only is it one of the easiest methods to implement (as we saw in *Chapter 6*), but it's also one of the most attractive from a computation standpoint. It requires no more memory than a standard neural network and, as we saw in *Chapter 7*, it's also one of the fastest models for running inference. These practical factors often weigh in more heavily than considerations such as uncertainty quality when it comes to selecting models.

Practical considerations are again likely the reason behind the second most popular method being deep ensembles. While this may not be the most efficient method in terms of training time, it's often the speed of inference that counts the most: and again, looking back to *Chapter 7*'s results, we see that the ensemble excels here, despite needing to run inference on multiple different networks.

Deep ensembles often strike a good balance between ease of implementation and theoretical considerations: as discussed in *Chapter 6*, ensembling is a powerful tool within ML, and so it's no surprise that NN ensembles perform well and often produce well-calibrated uncertainty estimates.

The last two methods, taking third and fourth place respectively, are BBB and PBP. While BBB is far easier to implement than PBP, the fact that it requires some probabilistic components often means that – while in many cases it may be the best tool for the job – machine learning engineers may not be aware of it or aren't comfortable implementing it. PBP takes this to a further extreme: as we saw in *Chapter 5*, implementing PBP is not a straightforward task. At the time of writing, there are no deep learning frameworks that incorporate an easy-to-use and well-optimized implementation of PBP, and – apart from the BDL community – many machine learning researchers and practitioners just aren't aware of its existence, as is made clear by the fact that it has fewer citations (although it still has a fairly impressive citation count!).

This analysis of the popularity of Bayesian deep learning methods seems to tell a pretty clear story: BNN approaches are chosen based primarily on the ease of implementation. Indeed, there is a significant amount of literature that discusses the use of BDL methods for their uncertainty estimates without considering the *quality* of the model uncertainty estimates. Fortunately, with the increasing popularity of uncertainty-aware methods, this trend is beginning to decline, and we hope that this book has equipped you with the necessary tools to allow you to be more principled in your selection of BNN methods. Irrespective of the methods used or how they're selected, it's clear that machine learning researchers and practitioners are increasingly interested in BDL approaches – so what are these methods being used for? Let's take a look.

9.2   How are BDL methods being applied to solve real-world problems?

Just as deep learning is having an impact on a diverse variety of application domains, BDL is becoming an increasingly important tool, particularly where large amounts of data are being used within safety-critical or mission-critical systems. In these cases – as is the case for most real-world applications – being able to quantify when models "know they don't know" is crucial to developing reliable and robust systems.

One significant application area for BDL is in safety-critical systems. In their 2019 paper titled *Safe Reinforcement Learning with Model Uncertainty Estimates*, Björn Lütjens *et al.* demonstrate that the use of BDL methods can produce safer behavior in collision-avoidance scenarios (the inspiration for our reinforcement learning example in *Chapter 8*).

Similarly, in the paper *Uncertainty-Aware Deep Learning for Safe Landing Site Selection*, authors Katharine Skinner *et al.* explore how Bayesian neural networks can be used for autonomous hazard detection for landing sites on planetary surfaces. This technology is crucial for facilitating autonomous landing, and recently DNNs have demonstrated significant aptitude for this application. In their paper, Skinner *et al.* demonstrate that the use of uncertainty-aware models can improve the selection of safe landing sites, and even make it possible to select safe landing sites from sensor data with large amounts of noise. This is testament to BDL's capacity to improve both the *safety* and *robustness* of deep learning methods.

Given their rising popularity in safety-critical scenarios, it should be no surprise that Bayesian neural networks have also been adopted within medical applications. As we touched on in *Chapter 1*, deep learning has exhibited particularly strong performance in the field of medical imaging. However, in these sorts of critical applications, uncertainty quantification is crucial: technicians and diagnosticians need to be able to understand the margin of error associated with model predictions. In the paper *Towards Safe Deep Learning: Accurately Quantifying Biomarker Uncertainty in Neural Network Predictions,*

Zach Eaton-Rosen *et al.* applied BDL methods for quantifying biomarker uncertainty when using deep networks for tumor volume estimation. Their work demonstrates that Bayesian neural networks can be used to design deep learning systems with well-calibrated error bars. These high-quality uncertainty estimates are necessary for the safe clinical use of models based on deep networks, making BDL methods crucial when it comes to incorporating these models in diagnostic applications.

As technology advances, so does our ability to collect and organize data. This trend is turning a lot of "small data" problems into "big data" problems – which is no bad thing, as more data means we're able to learn much more about the underlying process generating the data. One such example is that of seismic monitoring: over recent years, there has been a significant increase in dense seismic monitoring networks. This is excellent from a monitoring standpoint: scientists now have more data than ever before, and are thus able to better understand and monitor geophysical processes. However, in order to do so, they also need to be able to learn from large amounts of high dimensional data.

In their paper, *Bayesian Deep Learning and Uncertainty Quantification Applied to Induced Seismicity Locations in the Groningen Gas Field in the Netherlands: What Do We Need for Safe AI?*, authors Chen Gu *et al.* tackle the problem of seismic monitoring of the Groningen gas reservoir. As they mention in the paper, while deep learning has been applied to many geophysical problems, the use of uncertainty-aware deep networks is rare. Their work demonstrates that Bayesian neural networks can successfully be applied to geophysical problems and, in the case of the Groningen gas reservoir, could be crucial from both a safety-critical *and* mission-critical standpoint. From the safety perspective, these methods can be used to leverage the vast amounts of data to develop models that can infer ground motion activity and be used for seismic early warning systems. From the mission-critical perspective, the same data can be ingested by these methods to produce models capable of reservoir production estimates.

In both cases, uncertainty quantification is key if these methods are going to be incorporated into any real-world systems, as the consequences for trusting incorrect predictions could be costly or even catastrophic.

These examples have given us some insight into how BDL is being applied in the real world. As with other machine learning solutions before them, we learn more about potential shortcomings as the methods are used in more and more diverse sets of applications. In the next section, we'll learn about some of the latest developments in the field, building on the core approaches covered in the book to develop increasingly robust BNN approximations.

9.3 Latest methods in BDL

In this book, we've introduced some of the core techniques used within BDL: Bayes by Backprop (BBB), Probabilistic Backpropagation (PBP), Monte-Carlo dropout (MC dropout), and deep ensembles. Many BNN approaches you'll encounter in the literature will be based on one of these methods, and having these under your belt provides you with a versatile toolbox of approaches for developing your own BDL solutions. However, as with all aspects of machine learning, the field of BDL is progressing rapidly, and new techniques are being developed on a regular basis. In this section, we'll explore a selection of recent developments from the field.

9.3.1 Combining MC dropout and deep ensembles

Why use just one Bayesian neural network technique when you could use two? This is exactly the approach taken by University of Edinburgh researchers Remus Pop and Patric Fulop in their paper, *Deep Ensemble Bayesian Active Learning: Addressing the Mode Collapse Issue in Monte Carlo Dropout via Ensembles*. In this work, Pop and Fulop describe the problem of using **active learning** to make deep learning methods feasible in applications for which labeling data is time consuming or costly. The issue here is that, as we've discussed previously, deep learning methods have proven to be incredibly successful across a range of medical imaging tasks. The issue is that this data needs to be carefully labeled, and for deep networks to achieve high levels of performance, they need *a lot* of this data.

As such, active learning has been proposed by machine learning researchers to automatically evaluate new data points and add them to the dataset, using **acquisition functions** to determine when new data should be added to the training set. Model uncertainty estimates are a key piece of puzzle: they provide a key measure of how new data points relate to the model's existing understanding of the domain. In their paper, Pop and Fulop demonstrate that a popular method for **Deep Bayesian Active Learning (DBAL)** has a key shortcoming: that of over-confidence stemming from the MC dropout models used in DBAL. In their paper, the authors address this through combining both deep ensembles and MC dropout in a single model. They demonstrate that the resulting model has better calibrated uncertainty estimates, thus correcting for the over-confident predictions exhibited by MC dropout. The resulting method, dubbed **deep ensemble Bayesian active learning**, provides a framework for robustly employing deep learning methods in applications for which data acquisition is difficult or expensive – demonstrating again how BDL is proving to be an important building block in deploying deep networks in the real world.

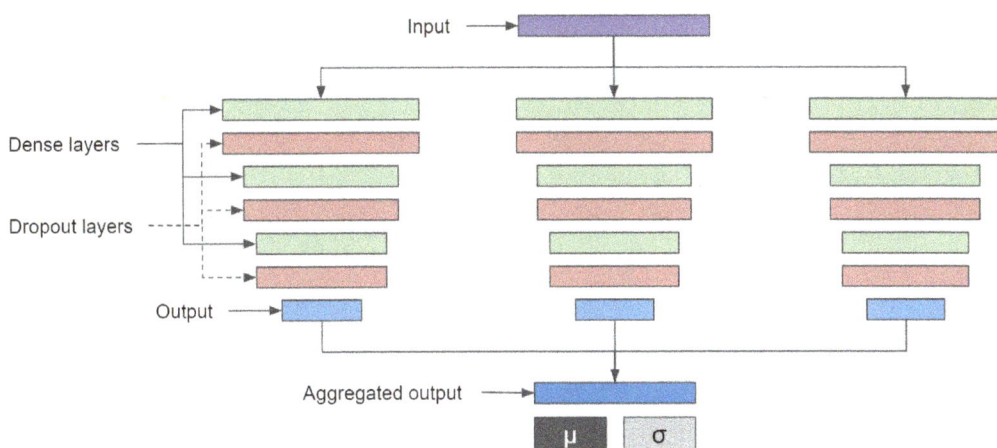

Figure 9.4: Illustration of a combined MC dropout and deep ensemble network

This approach of combining deep ensembles and MC dropout has also been applied in other applications. For example, the collision avoidance paper mentioned previously by Lütjens *et al.* also uses a combined MC dropout and deep ensemble network. This goes to show that it's not always simply a case of choosing one network over another – sometimes combining approaches is key to developing robust, better calibrated BDL solutions.

9.3.2   Improving deep ensembles by promoting diversity

As we saw earlier in the chapter, judging by the number of citations, deep ensembles are the second most popular of the key BDL techniques covered in this book. As such, it's no surprise that researchers have been investigating methods to improve on the standard implementation of deep ensembles.

In Tim Pearce *et al.*'s paper, *Uncertainty in Neural Networks: Approximately Bayesian Ensembling*, the authors highlight that the standard deep ensemble approach has been criticized for not being Bayesian and argue that the standard approach likely lacks diversity in many cases, thus producing a poorly descriptive posterior. In other words, deep ensembles often result in over-confident predictions due to a lack of diversity in the ensemble.

To remedy this, the authors propose a method they term **anchored ensembling**. Anchored ensembling, like deep ensembles, uses an ensemble of NNs. However, it uses a specially adapted loss function that penalizes the ensemble members' parameters from drifting too far from their initial values. Let's take a look:

$$Loss_j = \frac{1}{N}||\mathbf{y} - \hat{\mathbf{y}}||_2^2 + \frac{1}{N}||\Gamma^{\frac{1}{2}} \times (\theta_j - \theta_{anc,j})||_2^2 \tag{9.1}$$

Here, the $Loss_j$ is the loss computed for the jth network in the ensemble. We see a familiar loss in the equation in the form of $||\mathbf{y} - \hat{\mathbf{y}}||_2^2$. Γ is a diagonal regularization matrix, and θ_j are the parameters for the network. The key point here is the relationship between θ_j and the $\theta_{anc,j}$ variable. Here, the anc indicates the anchoring from which the method gets its name. These parameters, $\theta_{anc,j}$, are the set of initial parameters for the jth network. As such (as we see by the multiplication), if this value is large – in other words, if θ_j and $\theta_{anc,j}$ are significantly different – the loss will increase. Thus, this penalizes the networks in the ensemble if they deviate too far from their initial values, forcing them to find parameter values that minimize the first term in the equation while staying as close to their initial values as possible.

This is important because, if we use an initialization strategy that is more likely to produce a diverse set of initial parameter values, then maintaining that diversity will ensure that our ensemble comprises diverse networks after training. As the authors demonstrate in the paper, this diversity is key to producing principled uncertainty estimates: ensuring that the network predictions converge for regions of high data, and diverge in regions of low data, just as we saw in our GP examples from *Chapter 2*:

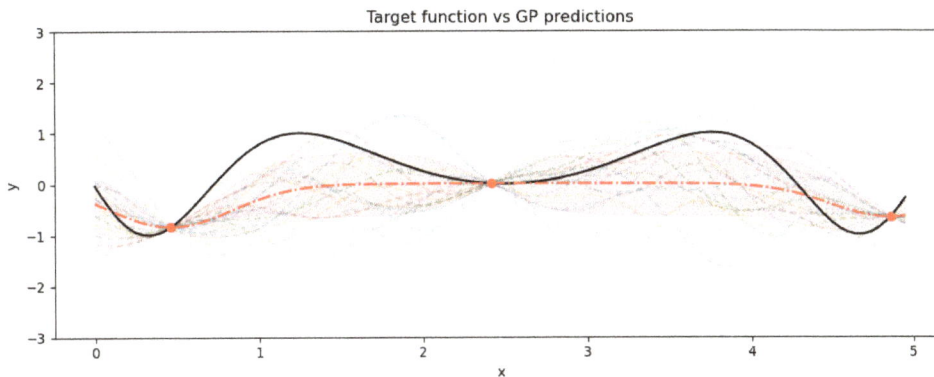

Figure 9.5: Illustration of principled uncertainty estimates obtained using a Gaussian process

As a reminder, here the solid line is the true function, the dots are the samples from the function, the dotted line are the mean GP predictions, the faint dotted lines are a sample of possible functions, and the shaded area is the uncertainty.

In their paper, Pearce *et al.* demonstrate that their anchored ensemble approach is able to approximate a descriptive posterior distribution such as this far more closely than the standard deep ensemble approach.

9.3.3   Uncertainty in very large networks

While the core aim topic of the book has been to introduce methods for approximating Bayesian inference in DNNs, we haven't addressed how this is applied to one of the most successful NN architecture varieties of recent years: the transformer. Transformers – just as more typical deep networks before them – have achieved landmark performance in a variety of tasks. While deep networks were already crunching large amounts of data, transformers take this to the next level: crunching enormous volumes of data, and comprising hundreds of billions of parameters. One of the most well-known transformer networks is GPT-3, a transformer developed by OpenAI that comprises over 175 billion parameters.

Transformers were first used in **Natural Language Processing** (**NLP**) tasks, and demonstrated that, through the use of self-attention and sufficient volumes of data, competitive performance can be achieved without the use of recurrent neural networks. This was an important step in NN architecture development: demonstrating that sequential context can be learned through self-attention and providing architectures capable of learning from hitherto unprecedented volumes of data.

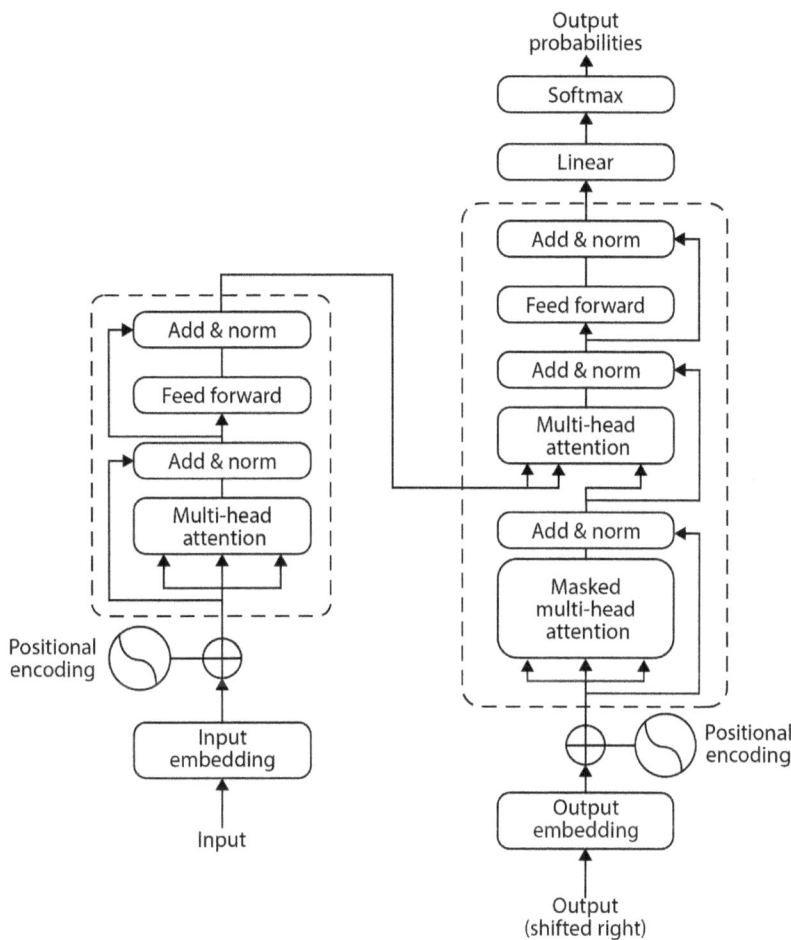

Figure 9.6: Illustration of the transformer architecture

However, just as with the trend of more typical deep networks before them, the parameters of transformers are point estimates, rather than distributions, thus preventing them from being used for uncertainty quantification. Authors Boyang Xue *et al.* sought to remedy this in their paper, *Bayesian Transformer Language Models for Speech Recognition.* In their work, they demonstrate the variational inference can be successfully applied to transformer models, facilitating approximate Bayesian inference. However, due to the large size of transformers, Bayesian parameter estimation for all parameters is incredibly expensive. As such, Xue *et al.* apply Bayesian estimation to a subset of model parameters, specifically the parameters in the feed-forward and multi-head self-attention modules. As we see from *Figure 9.6*, this excludes quite a few layers from the variational sampling process, thus saving compute cycles.

Another method proposed in the paper *Transformers Can Do Bayesian Inference*, by Samuel Müller *et al.*, approximates Bayesian inference by exploiting the large amount of data used to train transformers. In their approach, dubbed **Prior-Data Fitted Networks (PFNs)**, the authors restate the problem of posterior approximation as a supervised learning task. That is to say, rather than obtaining a distribution of predictions via sampling, their method learns to approximate the posterior predictive distribution directly from dataset samples.

Algorithm 1 PFN model training procedure

    Input: A prior distribution over datasets $p(D)$, from which samples can be drawn and the number of samples K to draw

    Output: A model $q\theta$ that will approximate the PPD Initialize the neural network $q\theta$

1: **for** i:=1 to 10 **do**

2:      Sample $D \cup (x_i, y_i)_{i=1}^m \approx p(D)$

3:      Compute stochastic loss approximation $\bar{l}_\theta = \sum_{i=1}^m (-\log q_\theta(y_i|x_i, D))$

4:      *Update parameters θ with stochastic gradient descent on $\nabla_\theta \bar{l}_\theta$*

As represented in the pseudo code here, during training, the model samples multiple subsets of data comprising inputs x and labels y. It then masks one of the labels and learns to make a probabilistic prediction for this label based on the other data points. This allows the PFN to do probabilistic inference in a single forward pass – similarly to what we saw in *Chapter 5* with PBP. While approximating Bayesian inference in a single forward pass is desirable for any application, this is even more valuable with transformers given their huge numbers of parameters – thus making the PFN approach described here particularly attractive.

Of course, transformers are popularly used within transfer learning contexts: using the rich feature embeddings from transformers as inputs to smaller, less computationally demanding networks. As such, perhaps the most obvious way to use transformers in a

Bayesian context would be to use its embeddings as an input to a BDL network – in fact, this is probably the most sensible first step in many cases.

In this section, we've explored some of the recent developments in BDL. All of these build on, and apply directly to, the methods we've introduced in the book, and you may want to consider implementing these when developing your own solutions for approximate Bayesian inference with deep networks. However, given the pace of research in machine learning, the list of improvements to Bayesian approximations is ever-growing, and we encourage you to explore the literature for yourself to discover the variety of ways in which researchers are learning to implement Bayesian inference at scale and with a variety of computational and theoretical advantages. That said, BDL isn't always the correct solution, and in the next section, we'll explore why.

9.4 Alternatives to Bayesian deep learning

While the focus of the book is on Bayesian inference with DNNs, these aren't always the best choice for the job. Generally speaking, they're a great choice when you have large amounts of high dimensional data. As we discussed in *Chapter 3* (and as you probably know), deep networks excel in these scenarios, and thus adapting them for Bayesian inference is a sensible choice. On the other hand, if you have small amounts of low-dimensional data (with tens of features, fewer than 10,000 data points), then you may be better off with more traditional, well-principled Bayesian inference, such as via sampling or GPs.

That said, there has been interest in scaling GPs, and the research community has developed GP-based methods that both scale to large amounts of data and are capable of complex non-linear transformations. In this section, we'll introduce these alternatives in case you wish to pursue them further.

9.4.1 Scalable Gaussian processes

At the beginning of the book, we introduced GPs and discussed why they are the gold standard when it comes to principled and reasonably computationally tractable uncertainty quantification in machine learning. Crucially, we talked about the limits of GPs: they become computationally infeasible with high-dimensional data or large amounts of data.

However, GPs are extremely powerful tools, and the machine learning community wasn't ready to give up on them. In *Chapter 2*, we discussed the key prohibitive factor in GP training and inference: inverting the covariance matrix. While methods exist for making this more computationally tractable (such as Cholesky decomposition), these methods only get us so far. As such, the key methods for making GPs scalable are termed *sparse GPs*, and they looks to solve the problem of intractable GP training by modifying the covariance matrix via sparse GP approximation. Simply put, if we can shrink or simplify the covariance matrix (for example, by reducing the number of data points), we can make the inversion of the covariance matrix tractable, and thus make GP training and inference tractable.

One of the most popular approaches for this was introduced in the paper, *Sparse Gaussian Processes using Pseudo-Inputs* by Edward Snelson and Zoubin Ghahramani. As with other sparse GP approaches, the authors developed a method for tractable GPs that leverages large datasets. In the paper, the authors show that they can closely approximate training with the full dataset by using a subset of data: they effectively circumvent the problem of large data by turning a *large data* problem into a *small data* problem. However, doing so requires selecting an appropriate subset of data points, which the authors term **pseudo inputs**.

The authors achieve this using a joint optimization process that selects the subset of data M from the full set N while also optimizing the hyperparameters of the kernel. This optimization process essentially finds the subset of data points that can be used to best describe the overall data: we illustrate this in *Figure 9.7*.

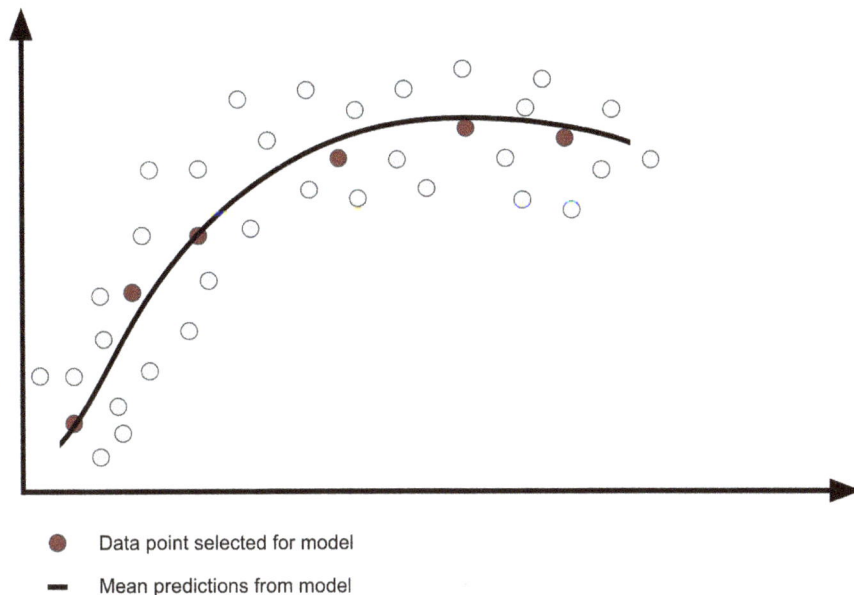

Figure 9.7: Simple illustration of pseudo inputs

In this diagram, all data points are illustrated, but we see that certain data points have been selected as they describe the key relationship between our variables. However, these points not only need to describe the relationship, in terms of mean, as a polynomial regression may do – they also need to replicate the *variance* in the underlying data, such that the GP still produces well-calibrated uncertainty estimates. That is to say, while the pseudo inputs effectively reduce the number of data points, the distribution of the pseudo inputs still needs to approximate that of the true inputs: if an area in the true data distribution is rich in data, thereby producing confident predictions in this region, this needs to be true for the pseudo inputs too.

Another method for scalable GPs was introduced more recently by Ke Wang *et al.* in their paper, *Exact Gaussian Processes on a Million Data Points*. In this work, the authors leverage recent developments in multi-GPU parallelization methods to implement scalable GPs. The method that enabled this is called **Blackbox Matrix-Matrix Multiplication (BBMM)**, reduces the problem of GP inference to iterations of matrix multiplication. In so doing, it makes the process easier to parallelize, as the matrix multiplications can be partitioned and distributed over multiple GPUs. The authors show that doing so reduces the memory requirement for GP training to $O(n)$ per GPU. This allows GPs to benefit from the kind of computational gains that have been benefiting deep learning methods for over a decade!

Both of the methods presented here do a great job of addressing the scalability issues faced by GPs. The second method is particularly impressive as it achieves exact GP inference, but it does require significant compute infrastructure. The pseudo-inputs method, on the other hand, is practical for a larger proportion of use cases. However, neither of these approaches tackle one of the key advantages of BDL: the ability of deep networks to learn rich embeddings through complex non-linear transformations.

9.4.2 Deep Gaussian processes

Introduced by Andreas Damianou and Neil Lawrence in their straightforwardly titled paper, *Deep Gaussian Processes*, deep GPs tackle the problem of rich embeddings through having layers of GPs, much in the same way that deep networks have multiple layers of neurons. Unlike the scalable GP work mentioned previously, deep GPs were motivated by the inverse of the scalability problem: how can we get the performance of deep networks with very little data?

Faced with this problem, and with the knowledge that GPs perform very well on small amounts of data, Damianou and Lawrence set out to see whether GPs could be layered to produce similarly rich embeddings.

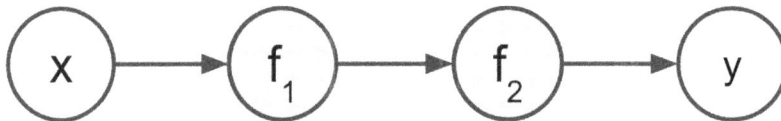

Figure 9.8: Illustration of a deep GP

Their approach, while complex in terms of implementation, is simple in principle: just as a DNN comprises many layers, each receiving an input from the layer before it and feeding its output into the layer after it, deep GPs also assume this form of graphical structure – as we see in *Figure 9.8*. Mathematically, just as with deep networks, a deep GP can be viewed as a composition of functions. The GP illustrated previously could thus be described as:

$$y = g(x) = f_2(f_1(x)) \qquad (9.2)$$

While this introduces the kind of rich non-linear transformations we're accustomed to in deep learning to GPs, it comes at a price. As we already know, standard GPs have limitations when it comes to scalability. Unfortunately for deep GPs, composing them in this way is analytically intractable. As such, Damianou and Lawrence had to find a tractable way of implementing deep GPs, and they did so using a tool that should now be familiar to you: variational approximation. Just as this forms an import building block for some of the BDL methods introduced in this book, it's also a key component in making deep GPs possible. In their paper, they show how deep GPs can be implemented with the help of variational approximations – making it possible not only to produce rich, non-linear embeddings with GPs but for rich, non-linear embeddings to be achieved with *small amounts of data*. This makes deep GPs an important tool in the arsenal of Bayesian methods and is thus a method worth bearing in mind going forward.

9.5   Your next steps in BDL

Throughout this chapter, we've concluded our introduction to BDL by taking a look at a variety of techniques that could help you to improve on the fundamental methods explored in the book. We've also taken a look at how the powerful gold-standard of Bayesian inference – the GP – can be adapted to tasks generally reserved for deep learning. While it is indeed possible to adapt GPs to these tasks, we also advise that it's generally easier and more practical to use the methods presented in this book, or methods derived from them. As always, it's up to you as the machine learning engineer to determine what is best for the task at hand, and we are confident that the material from the book will equip you well for the challenges ahead.

While this book provides you with the necessary fundamentals to get started, there's always more learn – particularly in such a rapidly moving field! In the next section, we'll provide a few key final recommendations, helping you to plan your next steps in learning about and applying BDL.

We hope that you've found this introduction to Bayesian deep learning to be comprehensive, practical, and enjoyable. Thank you for reading – we wish you success in exploring these methods further and applying them within your own machine learning solutions.

9.6   Further reading

The following reading recommendations are provided for those who wish to learn more about the recent methods presented in this chapter. These give a great insight into current challenges in the field, looking beyond Bayesian neural networks and into scalable Bayesian inference more generally:

- *Deep Ensemble Bayesian Active Learning*, Pop and Fulop: This paper demonstrates the advantages of combining deep ensembles with MC dropout to produce better-calibrated uncertainty estimates, as shown when applying their method to active learning tasks.

- *Uncertainty in Neural Networks: Approximately Bayesian Ensembling*, Pearce *et al.*: This paper introduces a simple and effective method for improving the performance of deep ensembles. The authors show that by promoting diversity through a simple adaptation to the loss function, the ensemble is able to produces better-calibrated uncertainty estimates.

- *Sparse Gaussian Processes Using Pseudo-Inputs*, Snelson and Gharamani: This paper introduces the concept of pseudo-input-based GPs, introducing a key method in scalable GP inference.

- *Exact Gaussian Processes on a Million Data Points*, Wang *et al.*: An important paper demonstrating that Gaussian Processes can benefit from developments in compute hardware through the use of BBMM, making exact GP inference possible for big data.

- *Deep Gaussian Processes*, Damianou and Lawrence: Introducing the concept of deep GPs, this paper demonstrates how GPs can be used to achieve complex non-linear transformations with datasets far smaller than those required for deep learning.

We've selected a few key resources to help you in your next steps into BDL, allowing you to dive deeper into the theory and helping you to get the most out of the content presented here:

- *Machine Learning: A Probabilistic Perspective*, Murphy: Released in 2012, this has since become one of the key texts on machine learning, presenting a well-principled approach to understanding all of the key methods within machine learning. The probabilistic angle of the book makes it a great addition to your collection of Bayesian literature.

- *Probabilistic Machine Learning: An Introduction*, Murphy: Another more recent Murphy text. Released in 2022, this is another important text providing a detailed treatment of probabilistic machine learning (including a section on Bayesian neural networks). While there is some overlap between this and Murphy's previous text, both are worth having at your disposal.

- *Gaussian Processes for Machine Learning*, Rasmussen and Williams: Perhaps the most important text on Gaussian Processes, this is a hugely valuable text when it comes to Bayesian inference. The authors' detailed explanation of GPs will give you a thorough understanding of this important piece of the Bayesian puzzle.

- *Bayesian Analysis with Python*, Martin: Covering all the fundamentals of Bayesian analysis, this title is an excellent piece of foundational literature and will help you to dive deeper into the fundamentals of Bayesian inference.

Index

‹packt›

www.packtpub.com

Subscribe to our online digital library for full access to over 7,000 books and videos, as well as industry leading tools to help you plan your personal development and advance your career. For more information, please visit our website.

Why subscribe?

- Spend less time learning and more time coding with practical eBooks and Videos from over 4,000 industry professionals
- Improve your learning with Skill Plans built especially for you
- Get a free eBook or video every month
- Fully searchable for easy access to vital information
- Copy and paste, print, and bookmark content

Did you know that Packt offers eBook versions of every book published, with PDF and ePub files available? You can upgrade to the eBook version at packtpub.com and as a print book customer, you are entitled to a discount on the eBook copy. Get in touch with us at customercare@packtpub.com for more details.

At www.packtpub.com, you can also read a collection of free technical articles, sign up for a range of free

Other Books You Might Enjoy

If you enjoyed this book, you may be interested in these other books by Packt:

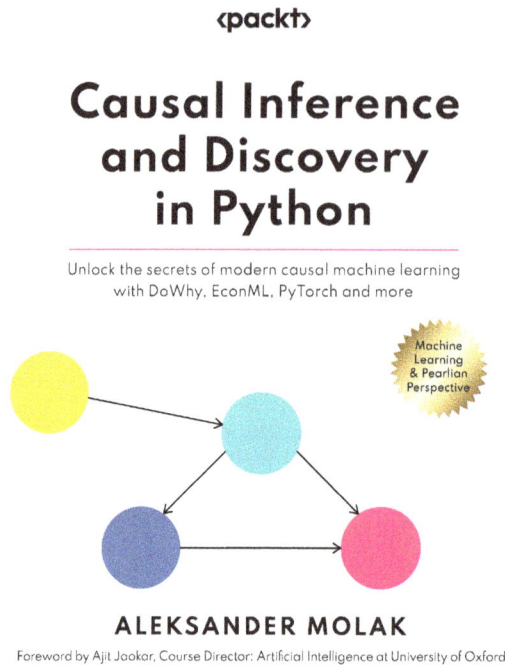

‹packt›

Causal Inference and Discovery in Python

Unlock the secrets of modern causal machine learning
with DoWhy, EconML, PyTorch and more

Machine
Learning
& Pearlian
Perspective

ALEKSANDER MOLAK

Foreword by Ajit Jaokar, Course Director: Artificial Intelligence at University of Oxford

Causal Inference and Discovery in Python

Aleksander Molak

ISBN: 978-1-80461-298-9

- Master the fundamental concepts of causal inference
- Decipher the mysteries of structural causal models
- Unleash the power of the 4-step causal inference process in Python
- Explore advanced uplift modeling techniques
- Unlock the secrets of modern causal discovery using Python
- Use causal inference for social impact and community benefit

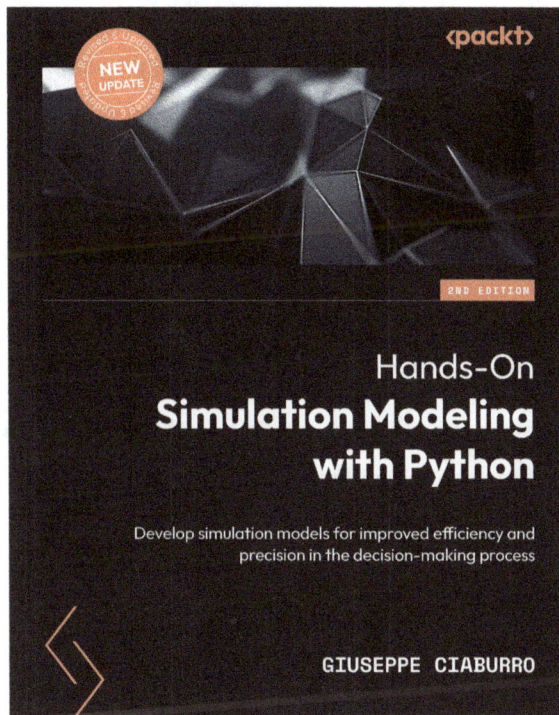

Hands-On Simulation Modeling with Python

Giuseppe Ciaburro

ISBN: 978-1-80461-688-8

- Get to grips with the concept of randomness and the data generation process
- Delve into resampling methods
- Discover how to work with Monte Carlo simulations
- Utilize simulations to improve or optimize systems
- Find out how to run efficient simulations to analyze real-world systems
- Understand how to simulate random walks using Markov chains

Packt is searching for authors like you

If you're interested in becoming an author for Packt, please visit `authors.packtpub.com` and apply today. We have worked with thousands of developers and tech professionals, just like you, to help them share their insight with the global tech community. You can make a general application, apply for a specific hot topic that we are recruiting an author for, or submit your own idea.

Share Your Thoughts

Now you've finished **Enhancing Deep Learning with Bayesian Inference**, we'd love to hear your thoughts! Scan the QR code below to go straight to the Amazon review page for this book and share your feedback.

https://packt.link/r/1-803-24688-X

Your review is important to us and the tech community and will help us make sure we're delivering excellent quality content.

Download a free PDF copy of this book

Thanks for purchasing this book!

Do you like to read on the go but are unable to carry your print books everywhere? Is your eBook purchase not compatible with the device of your choice?

Don't worry, now with every Packt book you get a DRM-free PDF version of that book at no cost.

Read anywhere, any place, on any device. Search, copy, and paste code from your favorite technical books directly into your application.

The perks don't stop there, you can get exclusive access to discounts, newsletters, and great free content in your inbox daily.

Follow these simple steps to get the benefits:

1. Scan the QR code or visit the link below:

https://packt.link/free-ebook/9781803246888

2. Submit your proof of purchase.

3. That's it! We'll send your free PDF and other benefits to your email directly.

www.ingramcontent.com/pod-product-compliance
Lightning Source LLC
Chambersburg PA
CBHW080708220326
41598CB00033B/5349